Teaching
Technolog

DEVELOPING SCIENCE AND TECHNOLOGY EDUCATION

Series Editor: Brian Woolnough
Department of Educational Studies, University of Oxford

Current titles:

John Eggleston: *Teaching Design and Technology*
Keith Postlethwaite: *Differentiated Science Teaching*
Jon Scaife and Jerry Wellington: *Information Technology in Science and Technology Education*
Joan Solomon: *Teaching Science, Technology and Society*
Clive Sutton: *Words, Science and Learning*

Titles in preparation:

David Layton: *Technology's Challenge to Science Education*
Michael Reiss: *Science Education for a Pluralist Society*

Teaching Science,
Technology and Society

JOAN SOLOMON

Open University Press
Buckingham · Philadelphia

Open University Press
Celtic Court
22 Ballmoor
Buckingham
MK18 1XW

and
1900 Frost Road, Suite 101
Bristol, PA 19007, USA

First Published 1993

A catalogue record of this book is available from the British Library

Library of Congress Cataloging-in-Publication Data

Solomon, Joan, 1932–
    Teaching science, technology, and society/Joan Solomon.
        p. cm. – (Developing science and technology education)
    Includes bibliographical references and index.
    ISBN 0–335–09953–X   ISBN 0–335–09952–1 (pbk.)
    1. Science – Study and teaching – Great Britain.   2. Science – Social
aspects – Study and teaching – Great Britain.   3. Technology – Social
aspects – Study and teaching – Great Britain.   I. Title.   II. Series.
Q183.4.G7S65   1992
303.48′3′071041 – dc20   92–8255
                                                                                            CIP

Typeset by Type Study, Scarborough
Printed in Great Britain by St Edmundsbury Press,
Bury St Edmunds, Suffolk

Contents

Series editor's preface

It may seem surprising that after three decades of curriculum innovation, and with the increasing provision of centralised National Curriculum, it is felt necessary to produce a series of books which encourage teachers and curriculum developers to continue to rethink how science and technology should be taught in schools. But teaching can never be merely the 'delivery' of someone else's 'given' curriculum. It is essentially a personal and professional business in which lively, thinking, enthusiastic teachers continue to analyse their own activities and mediate the curriculum framework to their students. If teachers ever cease to be critical of what they are doing then their teaching, and their students' learning, will become sterile.

There are still important questions which need to be addressed, questions which remain fundamental but the answers to which may vary according to the social conditions and educational priorities at a particular time.

What is the justification for teaching Science and Technology in our schools? For educational or vocational reasons? Providing Science and Technology for all, for future educated citizens, or to provide adequately prepared and motivated students to fulfil the industrial needs of the country? Will the same type of curriculum satisfactorily meet both needs or do we need a differentiated curriculum? In the past it has too readily been assumed that one type of science will meet all needs.

What should be the nature of Science and Technology in schools? It will need to develop both the methods and the content of the subject, the way a scientist or engineer works and the appropriate knowledge and understanding, but what is the relationship between the two? How does the student's explicit knowledge relate to investigational skill, how important is the student's tacit knowledge? In the past the holistic nature of scientific activity and the importance of affective factors such as commitment and enjoyment have been seriously undervalued in relation to the student's success.

And, of particular concern to this series, what is the relationship between Science and Technology? In some countries the scientific nature of technology and the technological aspects of science make the subjects a natural continuum. In others the curriculum structures have separated the two leaving the teachers to develop appropriate links. Underlying this series is the belief that Science and Technology have an important interdependence and thus many of the books will be appropriate to teachers of both Science and Technology.

Joan Solomon has been active and influential in the STS world since its conception. Having developed the seminal SiSCon in Schools course, she has continued to be one of its most stimulating advocates through her teaching, lecturing, examining and her research and development work. In this aspect of science teaching, she insists, the way that a teacher teaches, the way that a student learns, is of more fundamental importance than what is taught or learnt. Thus, in this book, she gives us insights into the processes whereby students develop their own thinking. STS teaching is not about knowing the applications of science in society, but about developing students' attitudes and about ways of looking at problems and deriving personally significant solutions to them. And it is about understanding the way that science works in its social context. This sensitive, perceptive insight into the way that STS works makes another important contribution to this developing field.

We hope that this book, and the series as a whole, will help many teachers to develop their science and technological education in ways that are both satisfying to themselves and stimulating to their students.

Brian E. Woolnough

Preface

Courses in tertiary level Science Technology and Society (STS) began to be taught systematically in Britain sometime between 1967 and 1970. By 1971, a group calling itself SISCON (Science In a Social CONtext) had been set up by Dr Bill Williams from Leeds University and was funded by the Nuffield Foundation. Like the school STS projects which started some years later, the founders of this movement never paused to define exactly what they meant by STS; they just designed course materials and got on with teaching it. This topsyturvy beginning did not seem to inhibit argument and discussion: indeed the caucus of university and polytechnic teachers who were involved met regularly to debate the issues, and for a while, the subject grew steadily. It was at the seminal SISCON Summer Schools in Coleg Harlech that I first learnt about STS and met its advocates. Clearly the 'subject' was being taught in different ways and for different purposes, so that a lack of formal definition seemed, for a while, to be almost more of an advantage than a hindrance.

In 1980 John Ziman's landmark book *Teaching and Learning about Science and Society* was published. Here he explored tertiary level STS, defining six different models of the role of science with respect to society. He listed, and largely rejected, a number of simplistic approaches to teaching STS which ranged from the narrow academicism of history and philosophy of science to a single-minded concentration on a particular issue – the problematique approach. It has proved an invaluable analysis by means of which the variety of possible purposes and approaches, but *not* any cut-and-dried definition, has gradually filled in the outlines of this elusive cross-disciplinary area.

Although the organisation of the present book, which is about school STS, is quite different from that of its predecessor, it does resemble it in one important respect. It too describes different approaches to STS in order to explore its nature. Once again the publication of teaching resources has run well ahead of any analysis of the subject. This sort of situation can continue for a while, but it is a sure sign of the growing maturity of a subject that its overgrowth begins to demand the construction of some clearer pathways. The enormously popular SATIS resources have given a higher than ever profile to STS, and the publication of some book about its teaching began to seem more and more of a necessity. This book may be just one possible answer to that pressure. It has proved a great help and advantage that, by the time of writing, some results of the recent research projects into the Public Understanding of Science were available, particularly those concerning the families of the youngest pupils and the discussion groups of much older students. This meant that the description of classroom teaching strategies could be filled out with some more analysis of their varied purposes at different stages in the learning and schooling of a child.

Three out of the seven chapters of this book describe aspects of STS in the normal fashion of an author – from the outside. Chapter 1 records something of the history of this approach both within the scientific movement and within education. This is valuable because it sets the scene by listing the areas of interest and the general aims that have been suggested for STS. Chapter 2 discusses what primary school STS might be like, and Chapter 7 records some results from research into how groups of adolescents talk about the social issues involved in STS. The whole of Chapter 4 is given over to descriptions of middle school classrooms where different types of STS lessons were going on.

However, it did not seem to me that this kind of writing could ever catch enough of the special flavour of this very committed type of teaching, nor of its variety. To do justice to that the actual words of teachers were required. So I wrote to, visited, and telephoned some of the most interesting STS teachers that I knew. With great patience and generosity they spent time explaining to me how they had introduced STS into their schools, why they had done so, what they found easy to teach and what was hard, and why they valued the enterprise. This material is to be found in the chapters about implementing STS in school, its special advantages for sixth-form teaching, and its use in examinations.

It is a pleasure to acknowledge my debt to the three main contributors. The first of these is Tony Hamaker whom I have known for many years and whose views and enthusiasm have had a great influence upon my own thinking and action. The other two, Sue Howes and Barbara Guy, I came to know later through their collaboration in the DISS project. Their special strengths and commitment to their students are explicit, different and enormously valuable. At two places in the text we also hear the voices of STS students whom I regret that I cannot thank by name. Further, acknowledgement is made to the Joint Matriculation Board for allowing access to confidential material from their Science Technology and Society examination scripts, presented in Chapter 5. The Joint Matriculation Board however, does not necessarily accept the views expressed in this chapter.

Finally, I commend this text to my colleagues, the science teachers for whom it was written. I trust very much that they will see within the text, despite its many drawbacks for which I take full responsibility, some particular items which they may find useful and even a small element which may inspire them to some new act of teaching or learning. They will know, as I do, that one's teaching can never stay the same, nor mark time. It is bound to change because our concern with science, with education, and with the students who pass through our classes, continually demand fresh answers to fresh questions, and that through the struggle to provide these our own understanding of science, and of people, goes on growing.

CHAPTER 1

What and why is STS?

How it all started

A neat way to begin this book would be to define just what STS (Science, Technology and Society) education is, or what it should be. In either case this would not be as easy a task as it might seem, since STS means different things to different people. This is largely because it has developed for a range of important but historically disconnected reasons. That makes one good reason for beginning with a brief look at its history. In the course of this we may begin to find answers to three questions:

● What have been the main themes in STS?
● Why should it be included in science education?
● Why does STS sometimes seem to be separate from, even antagonistic to, 'mainstream' science education?

There is some antagonism. Its severest critics see STS as a disreputable rival to 'valid' science. STS, they say, is concerned with the opinions of lay people, the arguments of politicians, the economics of profit-making, and the emotions of those who know little about science. It seems deliberately to search out the controversial and the topical and speaks about them in terms of passion rather than logic. On the other hand our traditional science, so they say, seeks quietly for eternal truths about Nature, using Nature's own incorruptible methods – disinterested experiment and incontrovertible mathematics. STS is in danger of corrupting the 'pure' science we have inherited from great scientists of the past.

Of course that description is no more truthful than any other caricature, but it does indicate the complexity of STS, which is concerned with lay people's troubles and views as well as with the theories and applications of science, and so does challenge its position. Caricatures sometimes make useful jumping-off points, but they shrug off any analysis of reasons and causes so that they make little contribution to our understanding. This one completely ignores the history of the STS movement, its purpose, and just why it includes so much which seems unscientific because it is related to the problems and politics of society.

In the first place it is clear that scientists themselves could never have been immune from such social influences in their own lives. All kinds and conditions of people have been scientists – those who were vain and self-seeking, and those who were intellectual recluses, those who were religious and caring, and those who struggled to make profit out of industrial innovations. And for each one of these, their science must have taken on a little of the flavour of the scientist involved because it is a human activity. In that sense at least science did not – and could not – stand aloof from the life of its times. That point always needs to be made to counteract the worst excesses of the ludicrous cartoon images of scientists that children's comics portray (see Chapter 2). But in the case of STS the whole rationale for its existence and study exists in the wider world which lies outside the laboratory.

'Contemplation and action'

When the foundations of British science were being laid down in the seventeenth century, there was a vigorous attempt to make science a part of the instrumentation of the state. In effect, science was going to be politicised. The originator of this movement was the illustrious Francis Bacon, Lord Chancellor of England and founding spirit of the Royal Society. In his book on *The Advancement of Learning* Bacon wrote with scorn about the kind of academic education he had received, common at that time, as 'no more than a conducted tour through the portrait gallery of the Ancients'. Bacon dedicated this book (perhaps the earliest ever to be written on the virtues of one kind of STS education) to King James I, and was astute enough to commend science as being useful to the state. The preface and much of the text is too fulsome for a modern taste, but there is no mistaking the tenor of the argument. He scoffs at the idea that learned men (scientists) should live a life of remote contemplation unsullied with civic matters – 'learned men forgotten in states and not living in the eyes of men are like images of Cassius and Brutus . . .'

> But the greatest error of all the rest is the mistaking or misplacing of the farthest end of knowledge . . . [which is] sincerely to give a true account of their gift of reason *to the benefit and gift of men*. This is that which will indeed dignify and exult knowledge, if *contemplation* and *action* be more nearly and straitly conjoined and united together. (p. 35, emphases added)

There was nothing new in recommending the applications of science to the rulers of states, and not least among the motives for doing this were the incentives of patronage and funding. These very human traits and needs are timeless, as many who are struggling to get grants for research know to this day. Galileo had offered his newly constructed telescope to the ruler of Venice as a valuable military device for observing enemy warships at a distance. He had an extended family to support and needed a rise in salary – which he duly got from his delighted employer. But Galileo was also committed to free contemplation, and he was later to argue fiercely against those of the cardinals who were attempting to channel his intellectual observation and speculation, that the evidence for scientific contemplation was there to be seen and read in 'the open book of the heavens'.

In Bacon's programme for science, the perspective was broad. He was claiming that there was an unfilled place waiting for the scientific enterprise, and for education in it, within policy and action on behalf of the state. Some time later when Thomas Spratt, Historian of the Royal Society, wrote to James II on the same subject he brought up the Plague and the Great Fire of London as the sort of public disasters which action based upon science might have been able to avert. This kind of agonised response to immediate disaster, or topical problems, is very familiar to us all. It calls for scientific action, and the call comes from lay people in need, not from scientists.

Bacon had been shrewd enough to argue for both contemplation and action, or, as we might put it now, for pure science and applied. The balance between these two was to characterise the swings from so-called 'valid' science, to STS, and vice versa, in the centuries that followed, and so eventually, to the educational curricula of the present age.

Even in the newly founded Royal Society, Bacon's practical and political aims for science were not embraced by the majority of the Fellows. The outstanding British genius of the seventeenth century was by nature more of a contemplative than a man of action or politics. When Isaac Newton's great work *Principia Mathematica* was presented to James II, probably one of the cleverest and best educated of all our monarchs, the King is said to have remarked about the dedicated volume that ' . . . like the Peace of God it passeth all understanding!' And that remark did Newton's reputation no harm at all in most circles. It was considered then, as it is today, stereotypically correct to be quite unfathomably clever if you are a scientist! Valid science still seems to some to be all about intellectual contemplation.

The knowledgeable and the workers

Royal or governmental patronage on the grounds

of utility did not link very well with contemplative science, even if the argument was that there might be useful spin-off for the state. While science remained an aristocratic pursuit, which it mostly was for the next three centuries, it did not need to chase economic return. This class system of science was a reflection of the contemporary culture, but it was to do considerable harm to the scientific enterprise in several ways. Robert Hooke, for example, was from lower class origins and spent much of his early scientific career making air pumps and other mechanical devices for more upper crust scientists, like Robert Boyle. This had unfortunate psychological effects on Hooke, but its effect on science had longer-term consequences. It divided the making of instruments and machinery from the thinking about them, just as it divided the population into those who might advance scientific thought, and those who would 'only' make technical devices.

Science suffered from the separation between making and thinking – between technology and abstraction. In the eighteenth century the inspired engineers of steam power were at a great remove from education in science, or indeed from any education at all. Some of the innovative engineers, like the Cornishman Richard Trevithic, could not even sign their names to patent forms. The abstract theory of thermodynamics arose out of the contemplations of a French savant, Sadi Carnot, more than a century after the clanking engines had been set to work. If it had been a class system that had divorced contemplation from action, then it had clearly been responsible for holding up the progress of science itself.

The separation also held back other kinds of progress. The Industrial Revolution produced pollution on a scale we can hardly imagine today, and there was no one to use scientific knowledge for combating it. By the middle of the nineteenth century the life expectancy in Manchester was a mere 28 years! There was almost no comment on social issues from the learned scientists, most of whom had not contributed to the new mechanical advances, and no recourse for those ignorant of science who had to operate the machines through the dangerous processes of trial and error.

By the end of the nineteenth century, there would be universal education, but even then there was precious little science in what was being taught. The question that the early twentieth century had to address was whether the public – society as a whole – had any role to play in science, and whether science as system of thinking was sufficient in itself to give the ordinary citizen, who had no intention of becoming an academic scientist, something of peculiar or personal value which should be included in their education.

Science for the people

The 1930s were a time when it began to seem that science might be reaching out to offer the citizen, and especially the poor and oppressed citizen, a prize of great value. J. D. Bernal, for example, claimed that he had come to realise by the age of ten, that science might be the saviour of his suffering countrymen in Ireland. He wrote in his autobiography (quoted by Werskey, p. 71) that:

> . . . the relief of the sufferings of the country and the possibilities of science, could in some sense be united. Science offered the means, perhaps the only means, by which the people of Ireland could liberate themselves. I saw it narrowly in a purely nationalistic sense, but it led me to an interest in science which grew to be a dominating interest in my life.

Lancelot Hogben, another member of this group of radical scientists, also tried to 'bring science to the people'. Like Bernal, Hogben had no doubts about the validity of what he offered. In the introduction to his book *Science for the Citizen*, Hogben wrote that it was intended 'for the large and growing number of intelligent adults who realise that the *Impact of Science on Society* is now the focus of a genuinely constructive social effort'.

So one STS educational objective was to deliver not the Baconian goods of science, but personal and social liberation through science. It was science for the people but, at this time, the science on offer had not been changed at all. Indeed for this illustrious group of scientists, science could take on this role precisely because of its high validity status. In ethical terms, as well as utopian and intellectual ones, they firmly believed that science had a morality which put it on the side of the oppressed.

> Science does not flatter our self-importance.

> . . . science makes stringent demands on our willingness to face uncomfortable views about the universe.

> Privilege is repelled by the [scientific] outlook because of its ethical impartiality.

For scientific humanists like these who were opposed to religion, it looked as if 'high science' served as a substitute morality. But this all happened before the Second World War which was to change the face of STS yet again.

The arguments of the humanists had seemed too political and avant-garde for most scientists in the 1930s. Although its new aims involved the public in science as never before, and for purposes of social betterment rather than wealth-creation, there was an essential ingredient of STS which was still missing. The radical scientists of the 1930s did not challenge scientific knowledge. So it is possible to see them as not so much the harbingers of a new approach to science and technology as a faction of the older unchanged 'valid' science order, despite their socialist ardour.

War and scientific responsibility

It is common to refer to the First World War as the 'Chemists' War' and to the Second World War as the 'Physicists' War', but the repercussions on science of the second were different in kind, as well as in discipline, from those of the first. In 1918 it had seemed enough to fix the blame for gas warfare in the trenches on the German chemical industry. The giant Krupps chemical complex was dismantled in 1919 by the victorious allies. The scientists who had originally developed those murderous gases which left thousands of ex-soldiers with damaged lungs to die slowly in the decade after the war ended, were not blamed. Scientists, it could be said, had identified mustard gas and other new gases without having any warlike intention. Those who should bear the blame were the ones who put the outcomes of what

was still called 'disinterested valid science' to work against Allied troops. That was how Lancelot Hogben could still write, in the 1930s, about the superb neutrality of science.

After the Second World War, the situation was completely different, for at least two reasons. In the first place the best scientists from all over Europe had flocked to Los Alamos to work, quite knowingly, on what they called 'our bomb'. The effects of actually dropping the Atomic bomb on Japan shocked them. In the aftermath they realised that it had not been just duty or patriotism which had made them work so hard at constructing this new and devastating weapon, but the basic scientific challenge itself. 'Whenever a problem looks technically sweet', said Robert Oppenheimer who led the team making the Atomic bomb, 'there will always be scientists ready to work on it.' This suggested that the scientific enterprise itself might be to blame: had it been carried away by intellectual problem solving and ignored morality? In a phrase which has made scientists cringe for half a century, Oppenheimer said 'The physicists have known sin.' It could have been better put, but the basic idea was clear – science, not just its products but the scientific attitude of its practitioners – was responsible.

After the war there were new organisations of scientists. One was the international Pugwash Conference which commented publicly on those aspects of science relating to world affairs, and managed to keep a membership of top scientists from the East as well as the West through even the chilliest phases of the Cold War. In Britain there was a less elitist association called simply the British Society for Social Responsibility in Science. Its members were all scientists (as opposed to modern committees on, for example, Medical Ethics) but the notion of responsibility to the public was now a minority, but growing, concept.

Philosophical and sociological attack

The other incidents which have affected the image of science are harder to pin-point. There was the Nazi insistence on a distinction between Ayrian science and Jewish science, and the Lysenko affair in Communist Russia which looked as if the science of genetics could similarly be branded as being of one political complexion or another (Medvedev, 1969).

These incidents might have been rejected as mere aberrations, if it were not for a growing interest in the ways of thinking of other cultures. Up to this time the West had drawn a line between what they dignified as 'science', and other 'unscientific' lines of thought. They believed that the pure thought of contemplative science could transcend all racial and national boundaries – on condition that it was carried out in their agreed fashion. Thus Astronomy was science, but Astrology was not; Western medicines were scientific, herbal treatments were not. Clearly the folk sciences of other peoples stood as examples of their world views, just as Western science exemplified our own preferred world view. This posed a puzzling challenge to the old certainties. Since world views differed, could there not also be a variety of sciences?

The last blow to valid science came from the philosophy of science itself. After centuries of reflection on the nature of logical thought philosophy had begun to look inwards, rather in the manner of a social anthropologist looking at an alien culture. It studied what the community of scientists did and how they agreed upon theories; and from that the new philosophy began to see the knowledge of science as an agreed outcome produced by a group of communicating scientists.

Four strands made up the argument for a new approach to science:

1 The capacity of science to provide wealth and health for society (from the top down).
2 The will to bring craftsmen, and then the general public, towards an understanding of science.
3 The shock of science in warfare, and later in the environment, and so the need for values and responsibility within science.
4 A re-evaluation of the neutrality of scientific evidence and knowledge-making (scientific epistemology).

Most contemporary philosophers view the construction of scientific knowledge as a much more fallible and human affair than one of strict reliance on the 'verdict of Nature' through disinterested experiment or logical argument. That was not to say that experiment and deduction ceased to figure in science. But the design of experiments, as well as the results from them, could now be viewed through culturally tinted spectacles.

Reasons for STS in science education

In the post-war years, before the new STS courses began in schools, two very substantial areas of need for 'citizen science' emerged in education. One of these was a public awakening to the environmental effects of new technologies which first made itself felt strongly in the United States. In the 1950s and 1960s, the extent of pollution had begun to shock the whole nation. Books like Rachel Carson's *Silent Spring* (1962), which drew attention to the quiet slaughter of birds and butterflies, horrified many. The Great Lakes had been badly polluted by industry with the result that one had become completely lifeless. Environmental groups demanded to know more about the activities of science, and this was reflected in a new Freedom of Information Act in 1967 which is still tougher than any comparable legislation in the UK.

But is extra information enough? Don't people need real understanding of science too if they are to express an opinion on the quality of the environment? By the 1970s there was expression of a new reason and function for science education.

All people need some science education so that they can think, speak and act on those matters, related to science, which may affect their quality of living.

Another push towards a new kind of science education came indirectly from an influential report by a group of the world's top intellectuals, economists and businessmen in The Club of

Rome. This commissioned a report about future global developments which would produce 'more understanding of the world's predicament', and 'stimulate new attitudes and policies' to address them. The report, *The Limits to Growth* (Meadows *et al.*, 1972), started a debate which included items such as the exponential growth in fuel use (then), and the finite nature of the fossil fuel reserve, the world population explosion and its limited production of food. The report was widely read and discussed. Shortly afterwards the Energy Crisis of 1973–4 raised the agreed price of petroleum so that the cost of petrol at the pump quadrupled in a single year. These events stimulated pessimistic television programmes in a vein which became known as the 'Doomsday Syndrome'.

Science was keen to bring its knowledge to bear on world problems. In the 1960s a group of American plant breeders working in Mexico had developed strains of new 'miracle grains' – wheat and rice – which launched the Green Revolution. This was designed to enable the Third World to feed itself and seemed to be scoring a remarkable success. Some scientists were optimistic about solving other problems too. They focused attention on the new technologies of alternative and nuclear power; they placed the problems of the developing world on the agenda of science. If science and its technology were to be the saviours of the world by solving these problems, as some seemed to believe, then this might be a new area of responsibility for science. At the very least it provided a new goal for scientific applications.

At first the 'Limits to Growth' debate in school was confined to questions on Advanced level General Studies, or Oxford and Cambridge scholarship papers. It is not surprising therefore, to find that the first school scheme in Britain to be entirely devoted to STS – Science in Society – was generated in public schools, for the sixth-form age group, and was particularly influenced by this report. (Problem-solving in a Third World context soon became a popular activity in the science lessons of some British schools largely, perhaps, because of the moral implications of 'doing good' for the unfortunate in other countries.)

> *Science education should address global problems, including those in the developing world.*

STS education begins

In general education there have always been different aims for students, and different perspectives on the purposes and problems of education. Somewhere among these STS, as it arrived on the school scene, would have to claim a place. In the 1970s this was not hard to do. Michael Young (1971), for example, examined the educational debate about 'equality of opportunity', and moved its focus from selection by ability (grammar school versus secondary modern), to the construction of curriculum itself. In science it was particularly easy to see the force of this argument: both traditional and Nuffield courses were about science for the intellectually elite. Later Malcolm Skilbeck (1984) wrote about the general purposes of education for 'social reconstruction' as well as for 'cultural transmission'. This new phrase implied that every generation has a right to act, within the constraints of the democratic process, in order to re-shape society. If an absence of scientific knowledge prevented people from thinking and acting on issues they cared about, then science education had seriously failed them. Both these educational arguments – about equality of opportunity within the curriculum, and about citizen empowerment – were tantamount to a new invitation. STS, with its emphasis on social responsibility, was needed inside the curriculum to complement the more traditional approach to science education.

Many educationalists thought then, and probably still do now, that just more and better science education of the same kind as before – valid science – was all that was needed. In Britain we had seen new schemes of the same kind in the Nuffield courses of the 1960s and 1970s. The founders of these talked about science as being 'clever thinking', and it was indeed in the tradition of contemplation and logic.

The first mainstream school science course to take up the challenge to include a strong STS component was SCISP (School Council Integrated Science Project, 1973). It satisfied several of the strands in STS which were set out in the previous section: it taught directly about economics and industry, as well as other STS themes. It achieved only a small following in the schools but arguably quite a large subsequent influence.

> *STS needs to teach about economics and industry.*

Educational progress rarely comes about in a logical progression, especially not in the old style British education system which had been (at least up until the Education Reform Act 1988) one of the most free in all the world. It was lightly driven by an examination system applied only at age 16. By the 1960s it had even become possible for individual school teachers to construct their own syllabus and examination for this age-group – the Certificate of Secondary Education.

One of the first school STS courses, SISCON-in-Schools, chose this simple way of breaking into schools and their examination structure, in 1981. The original SISCON (Science In a Social CONtext) project had been started in 1970 to promote STS teaching in universities and polytechnics. It produced teaching materials which linked science with economics in both the developed and developing countries, and others which taught the new studies of philosophy and sociology of science, or studied the connection between science and warfare. However while it still remained confined to degree courses, at least one strand of STS – the intention of bringing science to a wider audience so that the public could be empowered to take part in public and political issues – could not be realised. For that it needed to go into mainstream schools. In 1978, SISCON-in-Schools was begun by the few school teachers who had attended the parent organisation's summer schools. Like the earlier *Science in Society* this scheme was for sixth-formers, but it differed from that course in the strands of STS it chose to take up. Where the earlier course was strong on economics and industry, SISCON focused more on the human and

fallible nature of scientific theories, quality of life in the first and third worlds, and the social effects of new technologies including armaments.

> *STS should show pupils about the fallible nature of science.*
> *STS education includes discussion of democratic action.*

STS in other countries

Because societies differ, the complexion of their STS education will always have features peculiar to the educational system and the economic situation of the society concerned. There is no room in this chapter for more than a brief 'Cook's Tour' of examples pointing to those features which distinguish their use of STS from that in Britain.

Nations like the USA, which have always exerted more political influence on school science teaching, make their motives for change quite explicit. During the 1960s there were several reports on the nation's science education. The National Science Foundation wanted to increase the quality of science education so that there would be more scientists.

> *Education must transmit more scientific knowledge so that more of the new generation will want to become scientists.*

The government wanted more trained scientists to do scientific jobs so that the nation could match the Russians in new nuclear arms (or later to match the Japanese in industrial technology).

> *Education must teach science for technical expertise.*

These two objectives were exactly what a later British government would also claim as reasons for better science education – and it is noticeable that they nowhere mention any strong reasons for STS, nor, in general, did the new science courses include any.

Twenty years later, reports highlighted general confusion about educational goals for science education and linked this with public apathy towards science (Yager, 1980). From this time on STS began to figure prominently in discussion and in curricula (see McConnell, 1982). Historically the USA has always charged education in general, and social studies in particular, with developing the knowledge, attitudes and values conducive to 'good citizenship'. This was the most obvious niche for STS, and even today it is noticeable that topics like nuclear power and pollution are more often treated in social studies than in high-school science lessons.

One STS course notable for its advanced thinking appeared in Canada as early as 1971. Where provincial autonomy over education is so highly prized, gifted individuals can sometimes make great innovations. In *Science – A way of Knowing*, Glen Aikenhead produced a series of classroom materials for teaching the philosophical and social aspects of science within high schools which were in advance of those from any other country.

Holland had a special national need for STS in the 1970s which ensured that their new school courses had a matching objective. Several European countries had previously tried to solve the dilemma of whether or not to use nuclear power by means of a national referendum. Holland went further by setting aside a period of eight years for public education and consultation about nuclear power. It was during this time, and on the premise of public participation in decision-making, that the early Dutch STS courses such as PLON (Boeker, 1979) were begun.

Finally, the call for STS began to make itself felt in the developing countries. In almost all the ex-colonial nations there was an initial attempt to model school science on traditional and inappropriate Western courses. This was both because curricula materials were usually centrally designed and disseminated, and because the education of the more eminent scientists and educationalists in each country had been in Western universities. But borrowed educational themes and resources were often quite grossly at odds with local natural examples ('cows in the Oman'!), local culture ('British domestic plumbing in Africa'!), and local needs. No discussion of meaningful STS issues could flourish within such imported syllabuses.

Already new courses are emerging from the Third World which take STS issues in their own context (e.g. Health education in Child to Child, Bonati and Hawes, 1992) and teach school science around them. This movement may indeed produce personal and national liberation in something of the sense that the British scientific humanists had argued for, back in the 1930s, but of a different kind. Recently, a few Third World educationalists have also begun to consider the effects of cultural attitudes towards the environment in relation to school science courses (e.g. Jegede, 1991). Progress along these lines may begin to involve whole communities in a new kind of popular science which is deeply rooted in their own culture.

Summary

STS has, alas, proved too elusive to define, as was expected. However the origins and force of its varied purposes may have become a little clearer.

Special STS features within science education include:

- An understanding of the environmental threats, including global ones, to the quality of life.
- The economic and industrial aspects of technology.
- Some understanding of the fallible nature of science.
- Discussion of personal opinion and values, as well as democratic action.
- A multi-cultural dimension.

CHAPTER 2

Our youngest pupils

The social world of small children itself starts very small, and then it grows. But it does not only do this in the gradual way that children themselves grow, by the addition of other individuals who are introduced one by one to the child. Their social world also expands by sudden giant increases as the child enters whole new environments. One of these is the world of school; another is a vaguer entity, the world of television. Both of these play very important roles in the child's science education.

Learning about science, as opposed to learning science knowledge or skills, is an insidious affair. Children do not need to have science lessons, they hardly even need to hear the word 'science' mentioned, to begin to learn whether it is important or not, and in what way. Home is where the heart of science, like everything else, is. In the home children begin to form first impressions of science and scientists.

Growing up at home

Thinking about how a child in one's own culture comes to characterise science requires a little perspective. It may even be valuable to see what the social anthropologists who study child-rearing in different cultures have to say. Margaret Mead and Martha Wolfenstein (1955), for example, started their classic study of child-rearing with a list of 12 postulates or features which they believed were common to all cultures. Amongst these were the 'ideal adult character(s)' thought to be important in the culture, which influence how parents reward and punish their children. Expand the usual meanings of 'reward' and 'punish', imagine a scenario in which a parent is drawn to comment on an infant's curiosity about some natural phenomenon – 'A proper little scientist!' – and it is not hard to see, from the tone in which such a remark is made, how impressions may be created very early in childhood. Add to that another and totally unsurprising postulate by the same authors that 'habits established early in life influence all subsequent learning' and it becomes clear that the smallest child's image of science may not be trivial.

There is lamentably little in the way of empirical studies about children's very early steps in learning science, while whole libraries have been written about their acquisition of language and, to lesser extent, number. This is not just another plea for more or earlier science; it is a sober observation of the skew of interest and scholarship in a culture which is still predominantly literary and unscientific, a fact very relevant to STS. The influence of science within society is founded upon people's general 'feelings' towards science – such as familiarity, respect, suspicion or trust, every bit as much as it is upon the knowledge of scientific facts or theories.

Applying what we know about young children to the question of what they may first learn 'about' science, suggests that the societal aspects of science will not be at all easy to understand. It takes more than the whole first decade of a child's life to

Table 2.1 Number of children aged 5 to 11, in stages of societal thinking (percentages of age in parentheses)

Stage	5	6	7	8	9	10	11	All ages
I	15(94)	22(65)	2(8)	–	–	–	–	39
II	1(6)	11(32)	18(72)	23(76)	13(28)	5(15)	–	71
III	–	1(3)	5(20)	7(24)	32(70)	23(68)	7(64)	75
IV	–	–	–	–	1(2)	6(18)	4(36)	11

master the principles of the money basis of other people's jobs and roles, including the shop-keepers' necessary profit from buying and selling transactions, and the elementary functions of government (Stage IV, see Table 2.1). At earlier stages children still think of goods as being given to the shop-keeper by the factory, or as being made by the shop; and the government as a 'man with a wig'. Hans Furth (1980) carried out an extensive study of children's ideas through some delightfully described interviews about society and its trans-actions. Then he used a developmental approach to analyse the progress of children's thinking and to categorise it into four developmental stages of increasing complexity. In this he was quite con-sciously following in Piaget's cognitive tradition.

In Table 2.1, taken from Hans Furth's work, we can see that only 36 per cent of the children interviewed at age 11 had yet reached level IV. This makes sober reading for anyone anxious to include a strong STS element, of the conventional industrial top–down type mentioned in Chapter 1, at an early age. It is interesting to note that the only project developing STS materials for this age-group – 'Early SATIS (8–14)' – does include the involvement of industry in its 'Mission Statement', even though most of its resources, up to this point, wisely do not emphasise this very demanding strand of thinking. Without any proper under-standing of the economic basis of simple shop transactions children of primary school age are not likely to make very much sense of the role of science in industry.

There will be further requirements of social development for children if they are to study STS. In addition to some understanding of the world of work, our pupils, even the youngest, will need a developed sense of self and other people, and a sense of responsibility and justice which goes far beyond merely 'being a good child' (see Kohlberg's study of moral development; Kohlberg, 1984). Where Furth sees the child's development in all the social spheres as progress-ing through a series of Piagetian stages, with formal thinking using logical reasoning essential for performance at the top level for each one, other social psychologists have studied develop-ment from different perspectives. In particular the important work of Hartup (e.g. 'Children and their friends', 1978) is valuable because it crosses the primary/secondary divide and traces the de-velopment of the group collaboration and identifi-cation which comes into prominence during adolescence.

In STS education where empathy for another group is required, or role-play is used to establish what differing points of view are important on some technological issue, it does not seem likely that pupils of primary age will be ready to profit from its introduction. Such work needs a sure sense of self before empathy or role play can instruct or teach. On the other hand the complex notions of social cohesion and responsibility, which will later be drawn out through STS edu-cation, must rest upon a solid foundation of the simpler but basic moral and social notions con-structed during this early vital phase of education in the home and the primary school. To that extent at least, early education in social values is vital to later STS.

The world of television

This is a whole new social domain for little children – not just because they can get lost in its continuing soap operas and cult figures, nor even just because they may encounter those social issues which are properly a part of STS education in the occasional science programme; although that is important. Television forms a separate social world because it is the medium through which, unbeknownst to parents, children begin to experience out-of-home values and issues.

It used to be the case that children were introduced to the adult matters of society by their parents, learning about sex, or politics or employment, in the home as and when the parents thought it suitable. That is now vanishingly rare: much is learnt about these matters directly from television fiction. Neither is it the case, in most homes, that parents generally watch with their children and so are at least aware of what their children have been experiencing. Passage across this bridge from the ignorance of childhood to the knowingness of adulthood is no longer under the control of parents. Joseph Meyrovitz (1984, p. 28) put it this way:

> Socialisation can be thought of as a process of gradual, or staggered, exposure to social information. Children are slowly walked up the staircase of adult information one step at a time. . . . While young children once received virtually all of their information about the outside world from their parents, television now speaks directly to them, with the result that power relationships within the family are partially rearranged.

What effect does this information have on children if they receive it at an age when, as we have seen, social psychologists claim that they have such slight social maturity? It would not be surprising to find either that the children simply ignore all information that they cannot yet handle, or alternatively, that they are disturbed by it. No doubt both effects happen. The literature review of Barbara Tizard (1986), indicates not only the kinds of anxieties children have about social issues – including the threat of nuclear war – sometimes as early as the age of six, but also the parents' general ignorance about their anxiety. This is as might be expected if the children encountered

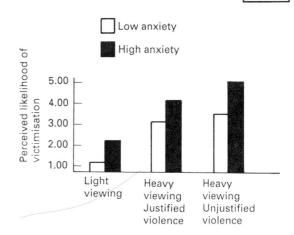

Fig. 2.1 Perceived likelihood of being a victim of violence as a function of amount and type of violence viewed in the Bryant, Carveth and Brown (1981) study *Note:* Based on data reported in 'Television viewing and anxiety: an experimental examination' by J. Bryant, R. A. Carveth and D. Brown (1981) *Journal of Communication*, 30, 106–19.

these complex issues within the separate and sometimes frightening world of television.

There is a heated and continuing debate about the effect that television programmes on sex and violence may have upon the young. This is outside the STS remit. However it is worth recalling the unfinished nature of this controversy, and also that reported anxiety is itself a social communication and may not actually describe any very troublesome behavioural symptoms. It is also interesting to find the general conclusion that, young children who are just beginning to make social sense of the world are more upset by chaotic and unjustified violence, than by dangers fully explained (Figure 2.1). This is an encouraging finding for the case of careful parenting, for school, and for STS issues. In the next section we shall find that such issues as 'the Greenhouse effect' and 'the hole in the Ozone layer' do have considerable currency amongst primary age pupils. Neither issue seems to upset them very much. (Of course that must not be taken to show that they fully understand any part of these issues. Even adults, or candidates for STS examinations at AS level, commonly attribute climatic change to the hole in the ozone layer, as though heat is leaking in!) Hodge and Tripp (1986) in their research into how children talk about their viewing conclude that (p. 214):

> Children . . . at least up to the age of twelve, not only prefer different kinds of programmes to adults, they also respond differently to programmes, and interpret them differently: but from the age of 9 they are capable of their own kind of understanding of most mainstream television.

As far as STS education from television is concerned we may conclude that most young children do watch programmes related to topical controversial issues, but without their parents' knowledge or connivance. Despite a basic lack of understanding about the commercial world of grown-ups, children above the age of nine may well be constructing, from their television viewing, their own simplified or distorted picture of scientists and science issues.

Children's images of science and scientists

Out of their various experiences children piece together their own images of the role of science and scientists in society. Because these images are constructed out of evidence from different circumstances and from different social worlds they may well be inconsistent. That does not mean, as we shall see, that one child may not employ several images at different times for different purposes.

One way to find out what children think scientists are like, and even what they do, is to ask them to draw a picture of a scientist. This has been done frequently from the classic work of Mead and Metrau (1957) to that of Williams (1990). One reason why this line of research has been so popular is, no doubt, that the data produced are far more striking than is verbal evidence from such young subjects. Another reason is the ease of its collection. Children seem to love drawing these figures and so, as at least one researcher has ingenuously commented, a great number may be generated with minimum effort on the part of the researcher. Two out of the recent collection of 11-year-old children's drawings made by Williams

are illustrated in Figure 2.2. The caption 'weird scientist' is really not necessary: the pictures show clearly enough their comic intention and origin. The most casual observation shows both the scientists to be male, bald, and engaged in chemistry, which may well have been dangerous and responsible for the bandages to both their faces! Williams assures us that the pupil artists who drew the pictures were in the first year of secondary school and had, thus far, done no chemistry at all in either primary or secondary school. They had never themselves used the foaming flasks which figure in their pictures. That marks out the pictures as stereotypes. We can confirm this view in two ways. In the first place the request to 'Draw a scientist' is tantamount to asking for fantasy rather than reality. It is as though one were to ask 'Draw a dragon'. The very wording of the instruction assumes that there is some figure which will have common currency as a cartoon with recognition tags on it. But it is also possible, by careful interviewing of the kind that Williams carried out, to show that the children themselves know that it is caricature. He asked the child artists if they had ever seen a scientist walking down the road.

> 'But I wouldn't recognise one if I saw him,' said one child.
> *When the interviewer pointed to the picture, the children all said*, 'They don't really look like that. That's just how you draw them.'

Cartoon figures may be known not to be 'real', but they are not altogether without power in any study of the images of science that children hold. They are generated by the world of films and comics, and we shall meet them again as children start to learn science.

Another way to probe young children's images of science is simply to talk to them. Williams and many other researchers have done this and the outcomes often show different images of scientists which may also derive from television viewing. Even though the more obvious idiocies of the cartoon scientist are now missing, gender stereotyping does not so easily disappear. Naima Browne and Carol Ross (1990, p. 49) have explored the gender stereotype amongst infants and nursery aged children.

Fig. 2.2 'Weird scientist'

ADULT: What is a scientist?

BOY 1: A scientist? I dunno.

BOY 2: It's a man.

BOY 3: It's a man, it's a doctor and he helps women and helps children. He gets his case and helps them.

GIRL 1: It's a doctor, a woman [*whispered*]. No it's a man [*said loudly*].

GIRL 2: It's someone who finds out what makes people sick. Testing things like yoghurt because hazelnut yoghurt got a poison in it and it makes people sick. [*This was a reference to a recent outbreak of botulism which was traced to hazelnut yoghurt.*] He tests water. It is a man. It's a man who tests medicines and make-up. He works in a school laboratory.

GIRL 3: You need science for doctors and nurses.

GIRL 4: It's like Doctor Who. They go into space.

In this short passage we have two or three images of the scientists in society which may be mixed together. One is drawn from current affairs' programmes on the media. The scientist tests things for us: he might be a doctor. Then there is a hint that the scientist is in a school, like a science teacher. At the end we have, once again, the 'weird' scientist.

Through interviews and responses to simple questionnaires which were carried out by our 'Nature of Science Project' (Solomon, 1991a) working out of Oxford University we also identified at least four distinct images, two of which have been already mentioned. In addition there is the scientist who makes things and who is said to work at improving them, and making them work better. This is the scientist/engineer and clearly has, like the testing and knowledgeable scientist on the

television news, something of value to contribute to society. Finally there is the scientist who is mostly interested in ideas. (There is a suggestion of the secondary school science curriculum in this image and indeed, amongst older children it is associated with the philosophical attitude of their science teacher.)

It was interesting to observe that, for each of these images of scientists at work, we found a slightly different rationale for experimentation. Only Image 2 seemed to be age-related, and more common among younger children. The others all flourished in the range from 9 to 14 years. It must be stressed, however, that children often identified more than one image. We interviewed them in small groups and so, in the manner of all social talk, the children often left one image and embraced another as they fell in with their friends' meanings and perspectives.

Referring back to Chapter 1 we can also identify, more or less precisely, some of the age-long diversity of purpose and practice within science in this range of children's imagery. The intellectual and the inventor are there, as is the explainer, who may be the scientist who saves society from disasters. (Even the children's comic caricature is to be found in the lampoons of Isaac Newton and fellow members of the Royal Society, written by Dean Swift in the eighteenth century.) One interesting outcome of our inquiries amongst young secondary school children was that the last of these images, the 'intellectual' in pursuit of corroborated theory, flourishes more in some secondary classrooms than in others, and seems to be related to the teacher's own philosophy of science. This point will be taken up in Chapter 4 during the discussion of STS material for this age range.

Scientist	Character	Purpose of work in society
Image 1	Weird	None. Dangerous and surprising experiments
Image 2	Authoritative and helpful	Tests and explains for us, like doctors and teachers (often said to explain Ozone Layer or Greenhouse Effect)
Image 3	Technologist (good?)	Makes artefacts for us, tests to see 'if they work' and improves them
Image 4	Intellectual	Has ideas, and tests them by designing experiments to see if predictions work

> Various different images of the role of scientists are co-existent in most children's talking and thinking. The media – comics, films and television – and the secondary school, may be the strongest influences in the construction of these images.

School/home links in science

So far we seem to have ignored the science that young children first learn in primary school. One reason for this is the splendidly 'everyday' character of most primary science. It is taught by non-scientists with the simplest of equipment and so manages to avoid a technocratic or esoteric aspect. This is, in itself, a sure foundation for the sort of 'citizen science' which STS seeks to teach. Later some of the simpler STS materials and strategies will be examined. These might be introduced into the lower secondary school classes with the express purpose of linking their school learning of science with society outside the school. For the primary school child, society is the family.

Many primary teachers set great store by the kind of class talk in 'Sharing time', or 'News time', by which they commonly make links with the home worlds of their pupils. One of the objectives of this work is to develop the child's vocabulary; it might also be to expand the pupils' everyday understanding to embrace school science, or to lead the findings of school science into their everyday world. These objectives are rarely stated, partly because the general aims of increased oral fluency and language development are, with good reason, the front-line interests of most primary teachers. But the argument from previous sections has emphasised the value of the social world of home for the young child. The effect of ignoring this mini-culture, and the images of science that it passes on to children, could quite well overwhelm the fragile edifice of primary science education.

Everyone knows that both home and school contribute to young children's general education. Perhaps it is not quite so easy to see how both can contribute to science education. Homes have no science apparatus, and many parents themselves have had little or no science education. A new project called SHIPS – School/Home Investigations In Primary Science – has been set up recently to build a bridge between home and school education (Solomon, 1991b). In the first stage of this project a playground survey of parents was carried out to see if parents in general, not just the middle class or knowledgeable, would be willing to help their children at home. Over 90 per cent said they would. Although the tally dropped to 70 per cent when the word 'science' was included, it rose again to over 80 per cent when they were told that this work would be a practical activity rather than more formal learning. This may be yet more evidence for the general frightening image of science amongst parents.

Two schools which trialled the work in the first year clearly agreed that it was essential that the activities were embedded in what might be called the 'culture of primary school science' as well as the culture of the home. So simple instruction sheets were written for science relating to the school's regular topics. ('Our community', or 'Transport' were quite easy to match with investigations; 'Festivals' was more difficult and 'Roman Britain' almost defeated the project's designer!)

The schools ran parents' meetings to explain what the science learning activities would be like and to answer their questions. These were small gatherings where two communities, both unfamiliar with the academic world of science, were bravely agreeing to work together in this somewhat alien world; it was, perhaps, not unlike the village meetings held to promote science or health activities in developing countries. The work has been running in different ways in each of the schools. One of the project team has been observing in some of the homes on a regular schedule during the year. Mothers, fathers and grandmothers were helping their children, each in their own way, make, talk and write about the science activity. This has been recorded and analysed to see how the science survives and adapts to its change of domain. In purely physical terms it is clear that the activities take on the character and culture of the home. Sometimes a space is specially cleared in a tidy room, on other occasions the activity takes place in a busy kitchen. Some parents try to take on the tone of a teacher, others just watch, comment and even marvel, alongside their children.

The home learning of science is not like learning in school. At home children have plenty of time and no classroom constraints or rivalry for attention. At home children ask questions, but these do not always get answered. Sometimes the child's comment is halfway between a question and the kind of slow reflection for which there is rarely enough time in the busy classroom. One child at home, for example, was floating a piece of Sellotape in water to see if it would cast a shadow on the base of the bowl. As she took it out another question arose in her mind and she wrote:

It is good to see the child so actively wondering and reflecting in this way, even if her query still hangs in the air. This is not academic science in which the uniquely correct answer is the only valid outcome of observation. Science cannot help but lose some of its 'valid' character when it enters the home, just as it does in all STS work.

When parents talk with their children the learning is easy-going and two-way, with each learning something from the other. In the following extract, Bob and his mother are looking at the impression of a coin that Bob has made on wax. His friend and his younger sister are also present.

MOTHER: You can see it's a coin actually can't you? because you can see the shape here, here. [*Bob reads the sheet for a little while. Then he gets up and goes to the cupboard.*]

yes I can see a
shadow – Black

I tried sellotape
and it made a
shadow but when
I took the sellotape
out of the water
it was still sticky
– why ?

writing Paper is
black shadow also
the Paper goes
soggy. I think
water can get
through this but it
does not get into
the foil. foil did

not go soggy.
writing paper is soft
foil is hard

When you hold
foil up to the
light it reflects
but you can't see
through it.

Paper does not
reflect and you can
see through it.

MOTHER:	What do you think is going to happen? Why have you got a different candle?
BOB:	It's thinner.
MOTHER:	What effect do you think that will have?
BOB:	That's thicker. It doesn't let the heat through. That one doesn't let the heat through.
SISTER:	One minute's gone.
BOB:	I'm going to leave it in for 2 minutes.
SISTER:	Why?
MOTHER:	This is his experiment. You can leave it in for longer . . .
BOB:	Oh, look at that! I told you it would be easier. Ah neat! You can see a shape on it.

No comment is needed on the pupil's ownership of science, or on the casual non-technical language which is used. School science learning is almost the opposite of this. Lateral wondering may be ignored, but skilful questioning from the teacher will draw children's attention to some important feature, compare their findings with what another has found out, or extend their use of language and understanding of science. In the extract below a class of juniors was discussing with their teacher a home activity on different kinds of fibres. The professional skill of the teacher is obvious: the link between science and society is more incidental and could be missed.

JOHN:	I picked some fluff off different carpets. Some were thicker.
TEACHER:	So there were a greater number of fibres. What about the fibres in the cotton wool?
JOHN:	When I felt them, they were more tight.
TEACHER:	They were more . . . ? What do we call it when they are packed tightly?
SALLY:	Compressed?
TEACHER:	Yes, compressed. A word like compressed?
LUCY:	Compact.
TEACHER:	Yes. So the fibres in this carpet were tightly compacted. And which material had more space in between?
GEMMA:	I got fluff from the corners in my room.

These kinds of activity allow science to spread into the home which is the young child's domain of society. Even after the activity is formally over,

there is evidence that some children extended the investigation in their own time over several days. Talk about it may surface while the family is at supper, or on holiday. This must be one of the surest ways to increase the *public understanding of science* and to give it the kind of familiarity at home which encourages children to go on speaking and speculating about it with their family. It fulfils, in its own small way, some of the basic objectives of STS listed in Chapter 1, such as the discussion of personal opinion within science, and a first weakening of its rigid epistemological claims which can be so forbidding.

If science is to be properly valued it must be able to fit into the culture of the home. Only the family itself can make this happen. Two examples from material collected by the SHIPS project serve to illustrate this. The first takes place in a do-it-yourself home where a boy is making a 'High-rise Crane' out of a Squeezy bottle in the kitchen. According to the instruction sheet he needs to add some damp soil to stop the bottle from falling over, but he has another idea. He goes out into the garden where the father is laying concrete and takes some back for ballast to put into his working model. Later, when some concrete spills out on to the kitchen table, the mother is commendably restrained – but then it is an acceptable occurrence in that home culture.

In the second example an infant, in school terms, has just made two rainfall measurers from jam jars and has tested them with the bathroom shower. The mother has talked with him about it and then wrote the following comment in the Parents' Diary: 'Olu was very excited and keen to complete this work. We discussed rainfall in the Caribbean and Nigeria – how important the rain is for these countries.' This demonstrates how such simple science ideas can be taken up and incorporated into the home culture of an ethnic minority. This begins to fulfil another of the objectives of STS in providing a multi-cultural dimension.

Summary

This chapter has made no suggestions for special STS course materials. It even argued that much of what we think of as STS material is out of reach for most young children. Instead the emphasis has been on describing the social worlds of the young child and images of science which derive from these.

There are two good reasons for this approach. In the first place it is possible, by constructing an atmosphere of natural acceptance of science in school and home, to build the foundations of a life-long attitude of familiarity and interest towards many aspects of science. This will stand the child in good stead in later life should some health or environmental problem threaten the quality of life, and some scientific knowledge becomes necessary for combating its real or imagined effects.

The second reason will be familiar to all good teachers. When any lesson has to be taught, the first stage in designing suitable materials and strategies is to 'ascertain where the student is', in the famous words of David Ausubel. The chapter has attempted to do that.

The only important and overt task being recommended in the name of STS, for which the young age of these pupils is particularly apt, is the forging of links between home and school. In this way science may become a talking point in both worlds of the child. It is easier to accomplish this before the onset of adolescence when children begin to outgrow the influence of home, and while the science being studied is still simple enough to take place in the kitchen or the garden. Such science will be concerned with the everyday objects that everyone uses. For STS, science should be set in the context of the significant people in one's society. At the beginning, these are to be found at home.

CHAPTER 3

Getting going on STS in the secondary school

'Models' of innovation

Changing from teaching 'valid' science of the type which lives in a world of its own on the textbook page, to teaching STS where the science is intertwined with technological response to individual needs and cultural values, is a big step. Somewhere along the line there has to be someone who has seen the world of teaching and learning through new eyes.

There are at least four types of agencies which can set out to produce changes in the science curriculum. These are listed in Table 3.1. Each one of them seems to have some effects, but in different places along the educational line. The agency for change is listed in the left-hand column and the locations of its likely effects are found in the four other columns. Some idea of how much change the agents are likely to produce is indicated in the order 'most', 'some', 'little' and 'none'.

This general outline does not imply, for example, that a change of government policy never

has an effect on the learning of our pupils. Of course it does, but not by itself. The National Curriculum for Science states that (Programme of Study at Key Stage 4):

> . . . pupils should be given opportunities to develop their knowledge and understanding of the ways in which scientific ideas change through time and how . . . the uses to which they are put are affected by the social, moral, spiritual and cultural contexts in which they are developed.

It is easy to see how extremely difficult that instruction is to implement as it stands. In countries where educational change has always been top–down, the government has learned that only if it effects changes in teachers will there be changes in the classroom. For this purpose they try to develop new curriculum materials and new INSET packages to make their policy change bear fruit for pupils. If teachers are influenced by INSET there is a chance that the proposed changes will take off in the classroom and so alter how children learn science. In Britain it had always been the task of

Table 3.1 Agencies to change the science curriculum

Agents	Location of change			
	Policy documents	Advice and INSET	Classroom teaching	Pupil learning
Government	Most	Some	None	None
Research	Little	Most	Little	None
Curriculum developers	None	Little	Most	Some
Teachers	None	None	Most	Most

Source: After Aikenhead (1989).

Local Authority advisers or inspectors to carry such policy decisions to both curriculum developers and teachers.

Much the same is true of the results obtained by educational research. There is much discontent in research circles about the lack of effect of the work they do. But researchers do not themselves teach, nor do they often produce curriculum materials which teachers can use for instruction. So the implications of their research are rarely implemented. It is well known in research circles, for example, that girls are more 'person orientated' with regard to what aspects of science interest them (Collins and Smithers 1984), and have less trust in technological progress (Breakwell et al., 1990), than boys. Both of these results have direct implications for science teaching in general and STS in particular, but they have had an almost negligible effect on teaching and learning.

Curriculum developers are usually more effective. Teachers can easily take on board the materials they produce, buying the books or photocopying the worksheets: SATIS (ASE, 1988) has been particularly successful in this respect. In so far as the content of instruction is concerned, this produces some obvious changes. But whether or not the learning process of children, and the image of science that they receive has changed, no one can yet be sure.

It is the teacher with a new idea about science education – with 'a bee in the bonnet' – who makes the real difference to how children learn science and what they think that this subject, science, is all about. One reason for this is that the decision to change is not only brought about by an ideological conversion, but also by an active dissatisfaction with the present teaching and learning situation.

Several different 'models' of change to STS teaching will figure in this chapter. Each one will focus upon teachers as the only effective change agents as far as children in the classroom are concerned. This does not mean that they act on their own without any regard to policy, advice, or new curriculum materials. On the contrary, as will be shown, both curriculum materials and the opinions of others are usually essential factors in successful change. But the teachers are the change agents because only their action can bring STS to the pupils.

Power to the secondary teachers?

The expertise of most primary teachers, as was shown in Chapter 2, lies in a professional understanding of young children and their struggles with language and expression. The secondary science teachers, by contrast, rest their understanding of children's learning on a substantial training in science. Scratch such a science teacher and you will likely find a science enthusiast. For what kind of science and science education are they so enthusiastic?

Choosing whether or not to teach STS might appear to be either a matter of curriculum choice, or of teaching strategy. In reality, the choice is more profound than either. It concerns the teachers' basic professional intentions and attitudes. The decision also brings to the surface some well-established attitudes towards science itself which teachers may have picked up during their own school or college education.

Britain is fortunate in having an all-graduate force of secondary teachers. It can count itself immensely advantaged that so many have graduated from an honours science course in a university or polytechnic. This academic background gives most of them considerable confidence in the substance of what they teach, which is illustrated in their authorship of so many good school science textbooks. This is by no means the case in all other developed countries, and gives some indication of the power that they are able to exert over secondary science education. It was practising science teachers, not teacher trainers or college professors, who wrote the first STS school courses, and it has been other science teachers who made the decisions whether or not to use them in their schools.

Even in the new era of the National Curriculum it is still the teachers who decide *how* science should be taught (Table 3.1), both its overall strategy and local classroom tactics. Examination committees and chairmen of examiners are still largely staffed by practising school teachers. The one major change of recent years was brought about not by curriculum fiat but by the size of comprehensive schools. It is now the team of teachers in the Science Department, rather than the individual, which decides on new developments.

This freedom of choice has always had opponents. In 1969, Musgrove and Taylor wrote rather extravagantly that: 'The freedom of teachers is the profession's glory: it is the people's shame.' The tension between the public and the professional has been sharpened to a ferocious edge by the current polemic about 'parent power' and 'teacher accountability'. More balanced dialogue between the two constituencies about the nature of the science that will be taught can do nothing but good for the pupils. However, it is only a Science Department that has discussed and decided upon its aims and intentions which is able to explain to parents the how and why of the particular course of science they propose to teach.

The next three sections will describe some of the different ways in which teachers have set about an innovation which led them towards STS. For most of this description the words will be those of the teachers, recorded while they spoke to the author, and transcribed verbatim.

Model I: Diagnosis and course material

In 1974, I think it was, I was at a school where we ear-marked the third year (14-year-olds) as the problem for that school. It was the option year and we were not getting kids to come forward and choose science. Our girls were not taking up physics and the kids in general were dissatisfied with science – that has all changed now. I wanted a course with personal and social issues because, as you know, I have always seen the need for that. R., my Science Adviser, suggested that we should start with the first year, start from the bottom and make a clean sweep, but we didn't have a problem with the first two years. There were no social issues in their course, you know, but the kids were enjoying it. It was all right. So we changed to SCISP for the third, fourth and fifth years.

That course was way ahead of its time. In fact,

later, when I first came back to teach in London and met my inspector at an ASE conference I launched into him. I said, 'You helped to produce this course. You were in on it at the start. It was ten years ahead of its time. Why aren't you pushing it in ILEA?'

I think he was rather taken aback by all this. Since then we have become good friends. I have a great deal of respect for him in terms of what his vision was, at that time, when the whole of STS began to take shape.

But the problem with SCISP was, in my opinion, that the books they produced were not very user-friendly and it was difficult for teachers to actually use them and get the main issues out. So, it is my guess, that for most of the teachers in that school, and in my present school when we used SCISP, that this made the course into one where they taught in the usual way as they had done before, and then just 'tagged on' the social issues at the end – if time permitted. In fact, to be honest, most teachers ignored the social issues to get at the 'true science concepts'. They did not integrate the social, economic and environmental issues with the 'true science'.

Remember when I say 'this is what we did,' I was not in other people's classrooms; I was teaching. I didn't know what they were actually doing, but I had a very good team of teachers and technicians then, and I have a good team now.

And I must say that the economic side of SCISP was almost a complete disaster. Essentially it was very difficult for kids to handle at Year 9 (third year). Looking back I think the problem was that those who developed SCISP did not have the research information that we have now about the psychological aspects, and the misconceptions. That came afterwards from the work of John Head [Head, 1983] about maturation and science choice, and also Michael Shayer's work on what children could understand [Shayer and Adey, 1980]. The people who developed SCISP just had a gut reaction about what might work.

At my school we adapted it, but the economic part was really difficult. You know in SCISP, all that time ago, there were short papers in German and French for the children to read. That's really cross-curricular, and I thought at the time, this is good. To be honest though those books were more boy-friendly than girl-friendly – but for the time, they were marvellous.

The above story was told by a Head of Science who saw a need in his school for a change to STS in one set of classes. He was already personally committed to STS, and picked up the only available course from the curriculum developers. Where it did not fit his needs, he adapted it.

It is easy to see how this story relates to the table of agencies for change on page – the Head of Science did talk over his plans with the Science Adviser, and yet it was his own professional concern with the uptake of science in his school which was the trigger for change.

He commented that the curriculum designers had made some mistakes (e.g. in economics) which he attributed to their lack of research information, and also, probably, to lack of teacher trialling.

Finally he guessed about the way the course had been taught by his colleagues and other teachers.

This illustrates the observation that new curriculum materials do not translate into changes in pupil learning unless all the teachers see themselves as agents for a change which they think important in itself.

Model II: All the Department decides upon change

We started it all with a brain-storming session to try to sort out, as science teachers, what we wanted to do. At that time, in 1983, we were teaching physics, chemistry, and biology separately. Nobody talked to each other: we each did our own thing. So at our brain-storming session we got each person to write down, on the board, just one aim of STS – Science, Technology and Society education. I remember they wrote:

Science is fun
Science relates to social, technical and economic issues
Science is about problem-solving

Physicists, chemists and biologists were writing the same things! You know the amazing thing was that not once during that two-and-a-half hour session did anyone say 'to teach physics', or 'to teach chemistry'. So what we did then was to prioritize those aims. We ended up with a complete consensus on eight global aims. The social, the techno-

logical, the environmental, the economic and the political were to be central to everything. And the problem-solving approach was central too. That was so wonderful for me!

We use 'problem-solving' in two ways in our work. The first of these is setting up scenarios for practical investigations. The team was interested in getting the kids to think about their thinking, so to some extent it pre-figured my interest in the CASE project [Adey and Shayer 1990]. It annoys me when teachers say, 'We are getting our children to think in science now.' We want to get them to think about their thinking in science – there is a real difference.

And secondly we use problem-solving in a social context – what is best for people. Here you are dealing with moral and ethical opinions, but you will find that very difficult to test in an end-of-term exam. SCISP tried to do that, I think. They used paper-and-pencil tests but these told us almost nothing about what the kids really thought – as we were to find out later in the DISS project [see Chapter 7]. Some of us tried to introduce discussion work but, I must admit, the teachers were rather blinkered about this. Written work was what we mostly did in place of discussion. But one thing we did find out – when we started to relate science to people the girls began to out-perform the boys!

Our teachers, which I inherited at this school, were very good and very experienced. After that brain-storming, everything was their decision. They were the change-agents. They decided to do SCISP in the third, fourth and fifth years, and SISCON in the sixth form, because they wanted to teach STS. And those two were the only courses which actually met our aims. Now we are moving over to Salter's Science for the fourth and fifth years, and again, when we feel that the social issues are missed out we use SATIS materials. Or, even more at this stage, we use STS materials we have written ourselves.

We have not agreed about particular strategies for the whole department. You see every time we did not agree on certain strategies other people did not carry them through. So now we are trying to be flexible, but we do all highlight in our Schemes of Work where the social, the technological, the economic and the political aspects can be brought in. These are still part of our central aims. They haven't changed since 1984 when we first wrote them down. These aims must *not* be rhetorical to be put in a drawer and locked away. Everything we do has got to be central to those aims.

What we are trying, in the sense of strategies, is to emphasise *discussion* and *working in groups*. In order to try and get pupils to discuss we give them folders and get them to highlight the issues they have discussed. (Some teachers are uncomfortable unless the pupils have something written down in their books.) At the end, one person from each group has to talk to the rest of the class about what they had discussed on that particular issue. We introduced the folders in order to monitor whether they were talking about the issue, or about last night's television, boy friends or girl friends! The idea was that they would use the folders to build up their own resources – something to look back to and reflect on.

For example, in Year 8 when we are doing acids and carbonates, as part of the scheme of work, they have got to consider 'the economic consider-ations of acid weathering'. We include the social aspects of acid rain, get them to discuss it, and then bring social questions into their end of unit exam-ination.

Basically we want our pupils to challenge, to take nothing at face value – not newspapers, not books, not even what I say – 'Let's collect and look at the evidence!'. That's the word I was looking for – evidence'. At fourth and fifth year they are still too young for economics and industry. They are using their own experiences and these are still a little narrow for industrial development. At this stage they cannot personalise industry or being a director. But we can teach them about collecting evidence. What we are trying to do is to develop skills so that when the children leave here, not necessarily being brilliant scientists, they can have the skills to be able to make judgements to en-hance their lives.

This Head of Science approaches change not by just importing a new course but by looking towards his colleagues to establish a common basis of shared aims. His comment 'That was wonderful for me' is a reaction from the 'first among equals' who now knows that his team can move forward together.

The whole Department then chooses what new courses to follow, on the basis of their established

aims. The common aims make it possible to trust colleagues' teaching without imposing common teaching strategies.

The STS aims inter-relate with other aims, such as fun for the pupils, problem-solving practical work, and thinking skills, which are not usually thought to be in the same domain. For this Department the connections between these sets of aims are so clear that they see no need to articulate them.

From the basis of common aims the Department can criticise the 'off the shelf' courses they have picked, and add to them materials they have collected from other schemes or, more often, written themselves.

The Schemes of Work instituted by the new National Curriculum are also used to forward their educational aims for STS.

The brain-stormed view about science merges into a wider view of education itself. From seeing the social implications of science, the teachers move towards seeing the function of science in the future adult lives of their pupils. This feeds back into further aims for their teaching in the class-room – stimulating the children to challenge statements and to use the evidence they have collected.

Model III: Teachers write curriculum materials

In the previous section the teacher went on to describe how he had developed new ways of teaching and new STS resource materials. Teach-ing strategies will be the subject of Chapter 4. This section will go on to describe another method of innovation in STS which is also teacher-centred. It is widely practised, having begun with the Science In Society course and continued by the SATIS team. The common factors have been a collabor-ation between

● teachers from different schools, and
● experts from outside education

for the purpose of writing resource materials.

There are other features which make this model very different from the previous two, such as substantial funding, a far wider audience, little or no debate about political issues, and more concentration on specific content related to industrial processes. It is the process of innovation – getting started on a new approach – which is the focus of the next two short case studies which are drawn from conversations with teachers who have been involved with one or other of the SATIS projects. The first of these began in 1984 when John Holman convened a weekend meeting of 30 science teachers. Since then it has recruited many more teachers all over England, and even abroad.

> I didn't start on the project because I knew a lot about STS, indeed I doubt if I had even heard those letters then. I was at a loose end, a bit in the doldrums with my teaching and getting stale I suppose. Someone asked me if I would like to join the local SATIS writing team because I had worked in industry before teaching, and I leapt at the opportunity. Joining the team was great – really inspiring.
>
> I wrote one unit almost completely by myself, but then it went out to be vetted by an expert. That happened to each one, I was told, when we had finished writing them. I met the person who had vetted mine and that was interesting too. I learned a lot more about how industrial research works than I had picked up from my own experience. They showed me round their outfit and later I spent the best part of a week there.
>
> As you can imagine that sort of thing has made me very keen to speak to my pupils about industrial processes when I am teaching. I do this quite often. I also use some of the other SATIS units quite regularly and they go down well. We have a whole filing cabinet of the sheets back at school, all photocopied and ready to use. I recommend the ones I know to my colleagues but I don't push them all that hard – you can't, you know – and I am not at all sure how much they are used. Peter and Sarah use them. They are also used by supply staff when teachers are ill, which is really very helpful.

This teacher has learned about one aspect of STS while already writing a unit. The organisation built on his existing strengths from experiences outside school, and expanded them. The teacher's work and attitude has been invigorated, but it has been less easy to pass that on to his colleagues who have not shared in the writing experience.

Model IV: Interstices model

> SATIS is not based on STS in the same sense as you were talking about. It starts, as John Holman says, with the school science syllabus – well, with a whole load of conventional syllabuses really in physics, chemistry, earth science, biology, astronomy – you name it! Then we look to see where there is a good opening for a SATIS unit. I think John calls it the 'interstices model'. That makes it very easy for teachers to use when they haven't got enough time to look for interesting extension material for themselves. Most people haven't got the time, have they?
>
> I have spent nearly twenty years teaching in several different schools, but I didn't teach any STS until SATIS came in. Before that I did a PhD in biochemistry. There was absolutely no STS implications in the work I was doing then. But I am able to use some of that basic knowledge for writing new units, and vetting others, on biotechnology for example. I have also provided 'balance' (you know?) when we get a rather way-out idea for a unit on the environment.
>
> It is enormously interesting work and I have learned a lot and spent some wonderful writing weekends in various places. There were experts there of different sorts, and I now know far more about the applications of science to society than I ever did before, even when I was still 'in science'. I have even learned a bit about scientific method and all that, which I suppose I should have known before, but I didn't.
>
> Yes I use some of the units regularly in my teaching. I like the simulation units very much. They are the best possible way of teaching our children what it is like to be involved in industry. They go down well; the kids learn to work in groups and present a case from the evidence given, both skills which I think are very important whether or not they become scientists when they leave. I am also very glad that the SATIS units have all been properly evaluated by asking teachers and students how they went. The results of that are superb – very encouraging.

This teacher, like the last, has learned about aspects of STS while working on the project. She has expert knowledge as well as teaching knowledge and uses both. It is interesting to observe that neither science teaching nor science research can be relied upon to generate reflection on the social, technological, philosophical or economic aspects of science.

The SATIS project does not attempt to make any special STS curriculum but uses traditional content and fits in isolated units of socially orientated material where they suit the content of existing syllabuses.

The materials are expected to recommend themselves to 'busy teachers', and are evaluated by both teachers and students in those schools which have used them.

Teaching and writing STS units has produced reflection on those skills that pupils may need as future citizens.

Summary

Innovations which start with classroom teachers, in whatever mode, usually work wonderfully well. Indeed there is a special term – 'the Hawthorne Effect' for this. Teachers generally work in isolation and when they do get together, whether it is in a brain-storming session in the Department, or in a writing weekend with others, it gives them a tremendous professional boost. If they are entrusted with the task of curriculum development in this context – they set about making it work with a will! But even if the reason for success does have more to do with the psychology of teachers than with the merit of the materials, that does not mean that this effect has no message for would-be innovators.

One suggestion is that all curriculum development should be based upon a large network of practising teachers. In the USA computers and databases have been used to create a network for STS (called NESTS) which runs across the whole country. In Britain, although we usually think smaller, our SATIS project has begun to develop a European network. In the models given above in the words of practising science teachers, it is clear that the small group which comprises no more than the science teachers in a particular school has special advantages in terms of mutual trust, implementation, feedback and Departmental objectives.

The small Departmental group also has some disadvantages. Chief amongst these is the lack of funding and expert support. Models A and B relied to some extent on knowledge culled from both INSET and LEA advice. The SATIS model, on the other hand, uses no background from the kind of INSET which gave these teachers access to research results about STS, the personal approach to science, or thinking skills.

To some extent the advantages and disadvantages of the two approaches may seem to balance out. However the complete tally of educational themes for STS, mentioned in Chapter 1 and briefly listed below, do make great demands on teacher understanding, and this presents a considerable problem for both attempts at innovation.

- Environmental threats to the quality of life
- Problems in the developing world
- Economic and industrial aspects of technology
- Teaching about the nature of science
- Decision-making skills
- Opinions on politically controversial issues
- Multi-cultural dimensions

Some of the above may be best coped with by expert advice on the SATIS model (e.g. industrial aspects, see Chapter 6) and others by cross-curricular collaboration within a school on the Departmental model (economics, developing world). The teacher in Model B, who had begun innovation with a brain-storming session, went on to mention his collaboration with Drama, History and Geography colleagues.

One last point about innovation concerns feedback information, to guide further change. Where a school has particular objectives for the new science course – such as in the Models A and B – this sets a marker for their success which can be used for monitoring and evaluation. Feedback from the classroom can then make suggestions for informal additions or deletions. This is true of any kind of school innovation and can be one of its most valuable features. The acid test for any vigorous STS course can be embodied in a single question:

Can the original innovation in STS be altered

subsequently to make it fit even better with the teacher's perceptions of new pupil needs?

If the answer is 'No' then the classroom teacher's power and involvement is unreal and will not survive. The innovation also can only age, harden, or be discarded.

CHAPTER 4

Teaching strategies in the secondary school

Choosing

The major STS themes were teased out in Chapter 1. They emerged from the history of science at work in the world, and from general considerations about education, but now need to be addressed in terms of classroom strategies. It is worth repeating the themes again here for use as a checklist for action. They were summarised at the end of the last chapter.

It is not enough just to list these, nor to comment upon the different ways in which STS teaching may be introduced into a school curriculum. The task in this chapter is to try to combine the teacher's aims, which are general prescriptions for outcomes in the classroom – 'science should be about people' or 'science should be fun', with the more abstract themes listed above. As they stand these are still at several removes from the classroom and the pupils. The task cannot be accomplished without offering a picture, drawn from school life, of how these teaching strategies work out in the classroom.

There is a problem with presenting teaching strategies in consecutive prose. Lists of resource materials will certainly not fit the bill, but neither will prescriptions for types of teaching without actual examples. In the belief that only vignettes of classes of learning pupils can show what some strategy actually looked like in action, this chapter runs the risk of falling between the two alternatives. It will draw on a specific example of a particular genre, so that a shortlist of resources

may seem to be presented. It will also make recommendations about how some strategies may be carried out which may seem too prescriptive. Each one of the following examples is securely based on personal experience.

Before the specifics are discussed in any detail there are two bogies to be laid to rest. *There is absolutely no conflict between teaching orthodox conceptual science for understanding, and teaching STS.* In the examples given below it is taken for granted that the pupils (aged 11–16) are engaged in learning science and technology in the normal curriculum. It would clearly be absurd first to teach something called 'science', or 'technology', in a way that made no contact with people and their lives, and then to clear the throat, as it were, and teach STS as though it had suddenly struck us that science had a personal, or a fallible dimension. In some books the impression is given that 'science' (whatever that is) must first be taught as an abstract discipline or an illustrated dictionary, and then the 'applications of science' will be tacked on afterwards. *The 'science' in STS education is school science and will be taught in just that spirit.*

Secondly, statistical data plus hand-waving talk is not the sum of STS. All science teaching requires practical illustration and activity. Environmental work needs field work, and other branches of science need the investigatory laboratory. Without a laboratory, science would become like gardening without a garden, or cooking without a stove. The whole feel of science, at least at the beginning, demands practical activities so that pupils can connect thinking about science with observing and exploring the actual phenomena of nature. Teaching which is all 'chalk and talk' may be inspiring in small doses if it is well done, but in too large an amount it distorts the very meaning of science. It becomes amassing information, instead of exploring natural phenomena. That means that any STS course must contain laboratory or field work.

Example 1   Using technology

The middle of STS, the 'T' for technology, is often lost in teaching for 'social issues' (Layton, 1988). Some argue that what is needed is no more than 'non-CDT technology'. By this they mean that it is not a matter of pupils designing and making, but only of discussing the benefits and risks of industrial innovation. That is the prevalent view in the USA, but not in Britain. Most advocates of STS here want real technological work. Time constraints are certainly very severe when children are set a task which involves all the blind alleys and craft skills essential to making something well. However technology is an experience that no child should either miss, or be led to believe is quite unconnected with science.

The most basic characteristic of all technology is that it is designed to serve people – society. It follows that 'egg-race' types of invention, although fun, forms no part of STS. In the following example technology leads the learning work, an approach which could easily be adapted for other topics.

Making a water turbine

The study of energy is a practical exploit in science for a number of reasons. In the first place young children believe energy to be something you can feel in your body, like 'being energetic'. There are ways in which such an approach, although formally incorrect, can be used. By making connections with the sensorimotor world of the child, as in Piagetian theory, energy may become simpler to understand.

However 'energeticness' is not the same as 'energy' because

- it cannot be measured
- it cannot be transferred to an object
- it cannot be stored

Good practical work should address all these points.

Secondly, energy, or the exploitation of it to do useful work for us, is at the heart of almost all technology. Here is where the social aspect of the work emerges. All pupils of secondary school age already have some idea that there is public controversy about nuclear power and about the emissions from coal-fired power stations. *Controversial topics are always most welcome in the classroom just because the pupils already know that the topic is thought to be important by people outside the closed world of the school.*

The children will know that there are energy technologies which rely on neither fuels nor nuclear power, like solar power or hydroelectric power. Some have learned that they are called 'renewable' forms of energy, but commonly interpret this to mean that 'the same energy is used again and again' (common response on STS examination scripts). These energy sources seem benign, 'natural', and non-threatening. What most people know little about, and need to know much more if they are to think and speak usefully about the subject, is how such energy is 'harnessed' and how efficiently we can use it. That is what the following exercise is all about (Figure 4.1).

The pupils must first think about how water-power has been used in the past for grinding corn, and how it is used in hydroelectric stations today, in order to give civic importance to their work. A few swift questions are usually enough to re-awaken some knowledge of the social context of water-power. The children will also probably know that it seems to cause no pollution, and think of it as 'good' technology.

They are then challenged to make a water turbine out of a cork, spindle, pieces of plastic (from yoghurt cups) which they can shape, and sleeves into which the completed rotor can be loosely fitted so that it turns easily. The turbine is powered by a flow of water supplied by the pupils raising and pouring a measured amount of water

Fig. 4.1 Making a water turbine and measuring its efficiency

(The general principle of this exercise can be applied to making a windmill, steam turbine, or chemical battery.)

You will need:

- A large bowl or deep tray
- Cork and spindle
- 2 glass tubes sealed at one end
- Thread
- Plastic cup
- Scissors
- Knife
- Funnel
- Rubber tubing
- Beaker
- A weight (about 200 g)

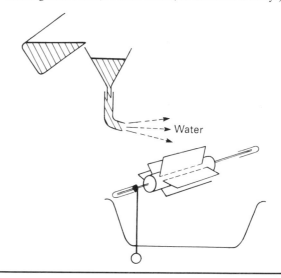

Water

(their energy, stored in raised water) on to the blades of the rotor.

The task of the turbine is to raise a weight.

So far this is a technological task: like all good technology teaching it starts with people's needs (power), and people's values (concern about pollution). The next stage is to improve the design – lengthening or broadening the blades of the rotor so that it turns faster or more powerfully for the same quantity of water flow. Later this will be translated into terms of energy. (With a little help the pupils can find out that 'faster' and more 'powerfully', are different and often competing conditions.)

Lastly, the pupils need to conceptualise the notions of energy and energy efficiency, which may have already started to form as they struggled to make a 'better' turbine than their neighbours. For this the pupils may be able to build on previous work. They will measure how much energy they have transferred to the turbine (weight of water × the height they have raised it). Then they measure how much useful work the turbine performs in terms of raising the small weight dangling on the end of the thread. The result is rarely more than 7 or 8 per cent.

There is no need to continue further with details of the exercise. As most teachers could tell from the account so far, this project has a slightly chaotic character. Water tends to spray around from some designs which spin too fast, or have very long blades. Carrying it out also has an element of competition. It is certainly fun.

This simple exercise provides a way to show students of almost any age that there is always a leakage of useless energy from a device which produces motive power (or electricity) for us. Both waves and wind are obvious sources of energy derived from the sun-driven atmospheric machine, but they are not quite the happy sources of endless power without any cost that some people seem to believe. They need excellent design, and manufacture which itself uses energy, and even then they will still convert only a small percentage of the available energy into electricity. The 'gradient' down which energy runs of its own accord with near 100 per cent efficiency is towards chaotic movement (swirling water, and heated substances), or chemical linkage, but not, alas, towards useful streamlined or rotary motion.

Either of the following make an enjoyable ending to this energy exercise:

- *pupils' strip cartoon* of how energy moves through the atmospheric machine which is powered by the sun's incoming heat energy, and is tapped for our use, or
- *stories and pictures for pupils* to show the risks or hazards of water power – the dreaded 'mill-race' of earlier times where the spent energy of the water could drown anyone who fell in – or photographs to show the environmental impact of a new hydroelectric dam, e.g. the Danube.

Water-power technology means, as technology always does, that people, like our students, have to make choices between energy, the environment, and human risk.

Example 2   Using history of science to understand its social relations and fallibility

In general conversation, and on the media, to say something is 'scientific' is tantamount to asserting it is 'entirely sure'. We may cast scorn on this between ourselves, but the problem of teaching science clearly to our pupils without giving this impression is substantial. Could we teach science at all if we really believed that all of it was 'up for grabs'? Classroom teachers need to explain authoritatively, and yet not give the impression that the explanation is certain for all time.

Two ways in which this may be done are through the interpretation of experiment, and through ideas from the history of science. There are plenty of experiments where the teacher can ask the pupils how they imagine what is happening, or explain what they have observed. This makes the point that scientific explanation does not just fall out of 'correct' experimental results. There is an uncertain pathway from experiment to theory along which it is quite essential that the imagination be used. Then it must be used again to make predictions for further experiments. (In a recent STS examination at GCSE level nearly half the

candidates gave an experimental result – like 'heavy objects fall at the same rate as light ones' – when they were asked to state a theory (Solomon, 1988). Another 25 per cent gave no theory at all.)

In the second context for teaching about the tentative nature of scientific explanation the teacher tells the class how the present accepted theory arrived on the scene. If this can be done so that pupils see that parts of the explanation are still missing today, then the point about fallibility is even better made.

Are there 'heat rays'?

(All objections to words like 'heat' or 'heat rays' have had a lower priority, in what follows, than the objective of making the concepts easily comprehensible to our young pupils.)

This work begins with a simple experiment, related to the pupils' own sensations, through which they begin to talk about how heat may travel from a heated wire gauze to their hands. The important discussion about how they interpret or imagine what is happening requires that groups of three or four pupils should work together. One or more of them put a patch of silver foil on the back of one hand, and a patch of thin black paper on the other. Holding both hands the same distance on either side of the heated gauze the pupils first find out that the black patch feels warmer.

Now they have to be encouraged to think, imagine and discuss what is happening. Children of all abilities find this equally difficult if they have never done it before – but all can do it. The teacher needs to find just one or two children prepared to break the ice, and then others will follow. The most frequent comments are:

'Black attracts heat'
'Silver repels heat'
'Black soaks up heat'
'Silver reflects heat'

The point to make about these alternative interpretations is not that one (the last?) is right, while the others are all wrong. Nor need we fight very hard to produce conceptual change. It is not the case that these interpretations are very tenaciously held by the pupils. Indeed a count of predictions made *before* the experiment is performed usually shows many more guesses that the silver will be hotter than the black one. Only when they have felt the hot black paper on their skin will the pupils begin to seek for possible explanations, sayings, or ways of imagining the result.

That process is not only the beginning of making meaning out of this experiment, it is also the first step in understanding that scientific theory is not 'discovered' by experiment, like a virgin continent might be by some intrepid explorer. In the terms that philosophers use, this approach avoids naive empiricism (assuming that the experimental result is the whole of the answer).

Designing the next phase of teaching requires a little reflection. Some teachers might want to proceed at once to teach the accepted modern theory. The problem with that approach is that it is tantamount to rejecting all the personal interpretations which we went to such pains to collect: 'What is the point', some pupils might then say, 'of risking looking a fool if s/he has the right answer up their sleeve all the time!'

Some constructivist researchers (e.g. Driver and Oldham, 1986) have argued that considerable time should be allowed for students to work out their own views in more detail. Still others reply that so many of the children's interpretations are only lightly held, that to concentrate on developing them runs the danger of fixing 'incorrect' views more firmly in the mind. The problem here is how to accord value to the children's views without labouring so much over the process that we actually reinforce them.

One resolution is to point out that diversity of interpretation is almost inevitable for any new observations, and has happened regularly in the history of science. By tracing some of these early guesses, and the findings which changed what scientists then believed, the teacher has an opportunity to demonstrate more about the nature of science in general, and about the properties of heat radiation in particular. It also treats the pupils' efforts to interpret and understand as an integral part of the scientific endeavour.

At this point the pupils need some historical information. When such passages are selected it is always important to remember that history itself can be presented in several ways. It can be

(a) conceptual and chronological, or
(b) related to scientists and triumphant, or
(c) social and imaginative.

From the first of these our pupils learn only how one idea followed another, and when this happened. From the second they learn stories of great men, and very occasionally women, doing better and better as time goes on! (It tends to reinforce the idea which many children hold that 'we are more intelligent nowadays'.) From the third kind of history they learn the basic stuff of all STS.

The problem for teachers is how best to introduce the historical information. It can be provided in a series of snippets, one for each group, so that each passage is not too long and complex, but this still does not answer the question of how we can best help the pupils to study it. No experienced teacher needs to be told that pupils do not just sit quietly reading and absorbing information! Probably the best way to proceed is to provide an activity which makes them search the text in some interesting way. These sort of activities are sometimes called DARTs (Directed Activities Related to Text) (Davies and Greene, 1984).

In the case of heat radiation it has proved successful to ask each group to prepare a poster which incorporates the information in their short passage. Using this they can then explain to the rest of the class their piece of the history of understanding about heat rays. The posters are then pinned up around the room for other classes to see as they enter the laboratory.

An alternative DART is making an overhead transparency. This can be excellent for a subsequent mini-lecture (2 minutes), so long as it does not just have words from the text on it.

An assortment of pieces of text which can be used are to be found in a unit in *Exploring the Nature of Science* (Solomon, 1991a, pp. 53–56). The examples given run as follows:

● An old idea was that heat and light from flames are particles.

- John Leslie finds that black surfaces of metal containers holding boiling water give off more heat than silver ones, but that the heat will not penetrate glass.
- Caroline Herschel finds that heat from the sun goes through some of the telescope filters which absorb light.
- William Herschel finds that heat lies beyond red in the visible spectrum. He argues that it is 'invisible light'.
- Photographs can be taken using heat rays. This is used for studying distant stars and galaxies, and also in war for detecting soldiers and engines.
- Heat radiations can be used for cooking or in medicine.
- Glass lets through radiation from the sun, but traps radiation from cooler objects. (This example leads to a discussion of the Greenhouse Effect.)

The pupil groups then deliver their own explanation. Experience shows that even shy pupils can make some attempt at facing the class and talking coherently when they are part of a group which is armed with a poster to point at.

So far this exercise has used both class experiment and the history of science, and has touched upon the uses of radiation in medicine and warfare. It has also brought in an environmental topic which has been much debated in the media – the Greenhouse Effect. STS teachers will want to discuss this with their pupils, and to hear what they think, at several points in their science curriculum. For the purposes of teaching about scientific explanation it is valuable because we can acknowledge our present lack of knowledge about the details of this effect.

Young children are apt to say that scientific theories are either 'facts' or 'guesses' rather than attempts to predict and explain. What we have taught them so far about heat rays begins to correct the first impression that science just relies on 'facts' (Solomon et al., 1992). However it has done less to teach them about prediction from a hypothesis. This is now remedied.

> The pupils are reminded of William Herschel's theory that 'Heat is like invisible light'. They are asked what they would expect heat to do if it really was like light. (This can be a group investigation where they discuss together, design an experiment, and then carry it out.)
>
> Boxes of mirrors, pieces of foil, lamps, lenses, thermometers, and heaters are put out for the pupils to use in their investigations. They also serve to trigger ideas about reflecting heat, making heat shadows, or focusing heat on to a thermometer. (If prisms are put out experience suggests that several groups will just attempt, and probably fail, to repeat William Herschel's experiment.)

Scientific theories, like this one of Herschel's, are often models for how things work. That means that they can be manipulated in the mind, almost like a mechanical model, to make a prediction for what may happen under new circumstances. Here is another important task for the imagination, in addition to the original act of personal interpretation. This special function of theory may take several lessons on different topics for the pupils to begin to understand it clearly, but a beginning is certainly possible in the simple manner of this exercise.

Example 3 Using the cultures of different peoples to deal with different local problems

Of all the objectives of STS in our list this is undoubtedly the one which is most likely to go horribly wrong. With the best of motives both the famine relief organisations and good-hearted teachers, give the impression that the poorer non-Western countries are populated by malnourished, uneducated people who have too many children. To use this image, even for the purpose of collecting money or stimulating good will, is indefensible. Our task in STS is not to offer charity, nor to examine other living conditions in order to devise yet more applications of simple school science. We study local problems in other communities in order to see how the people use or extend their indigenous technologies to protect their own style of living. Its objective is to observe and learn from other societies how they use technology to preserve or extend their quality of life. Like all STS it should be about real people, their objectives and their issues. Reading well-meaning answers to STS questions about Third World medicine shows how often this objective has been completely missed. Students too often impose our issues and ways of living (small families, settled living and western medicine).

The stone walls of Burkina Faso

The sub-Saharan countries have been growing more arid over the last centuries for at least two reasons. One is a climatic change. The other reason is the confining of previously semi-nomad farmers to these precarious dry regions from which they can no longer retreat to better watered southern lands in seasons of drought, since the latter are now enclosed and farmed. The semi-arid farms in this flat region always had a problem of holding back the precious rain water during the short, occasional, but heavy downpours. Now this problem has been made more severe since the land is in regular cultivation, so the soil is loose and can be washed away.

The old solution to this problem was to build lines of stones to hold back the water as long as possible to soak into the ground. Now it becomes even more essential that these lines follow the contours very exactly otherwise the flash flooding may be channelled by the stones and swirl along the walls taking even more soil with it. The technological question was how these farmers could use the old custom for greater benefit. Oxfam sent out scientists to consult with farmers and together they came up with a simple manometer made from plastic hose attached to two measuring sticks. With this they can now mark out level lines on this nearly flat landscape for laying down the stones far more exactly. The farmers have the task of finding and transporting the stones (Figure 4.2). This simple aid to existing practices can increase crops by at least as much as 35–40 per

The hosepipe is filled with water

Fig. 4.2   Equal water pressure

Pupils learn about air pressure and water pressure in the usual way, as African children will also learn it. Then they make a model of the manometer and use it to mark out the contours on some uneven piece of land. I can vouch for the fact that this requires group collaboration of a high order. A dropped stick at one end involves spilled water and may mean going back to refill the tubing, and even starting all over again.

To introduce the topic it is suggested in the book that groups of pupils once again read pieces of text and use an activity to illustrate their understanding of the information. This time drawing posters has drawbacks. Because the African context is so foreign to our pupils they draw pictures with considerable uncertainty. What they produce inevitably bears little resemblance to the actual landscape or plant-life of Burkina Faso. Where texts in other circumstances may be illustrated by simple or humorous cartoons, these are obviously less appropriate for cross-cultural work.

cent. The classroom resource is to be found in the same textbook, *Exploring the Nature of Science* (1991).

A local issue

The most valuable outcome of a cross-cultural study in an STS context is not a detailed understanding of some technological development in another country. It is a more general understanding that every locality has difficult issues, threats to soil fertility and problems with land boundaries. Each neighbourhood tries to cope with the issues using scientific and local knowledge about the environment, technological inventiveness, and concern for how people live. In Britain country children will know about the controversies over hedges, and town children about threats to parks from diseases spread by dog excrement. The soil is precious everywhere, only the issues are local.

A successful method of studying such issues is to use local newspapers. This has the advantage of being in a familiar setting so that our pupils are in a better position to understand the concerns of people. The disadvantage is that it is not possible to plan for some appropriate local crisis just as that point in the syllabus is reached! There are two strategies for dealing with this. One teacher may arrange to collect newspaper cuttings on problems affecting the neighbourhood over a few months and use them at an appropriate point in her scheme of work. Others may make an opening for this kind of exercise when the news breaks – it may be pollution of the water supply, trying to make a graffiti-free surface for bus-shelters, or looking after the elderly in some different way.

It is clearly impossible to dictate a soil-related issue which will fit neatly with the African example above: that is a disadvantage. On the other hand it will now be the pupils themselves who collect the information from newspapers, local radio, and people that they know, and that is clearly an advantage. Above all the issue will be real to them,

---

The pupils cut out newspaper articles and record radio interviews wherever possible. These are brought to school and the class activity is to mount them together with connecting passages which groups of pupils are asked to contribute. To do this the pupils are encouraged to talk in their groups while planning and writing so that they can exchange information and begin a discussion of the issue and what *they* think should be done. At the end there is a class display and also a plenary discussion with the teacher in which she draws out the reactions of the children and shows that she values their contributions.

---

topical, and about people they know. It will enable them to begin to form their own opinion on a civic issue, and even to see what action might follow to deal with it.

Example 4   Using role-play

The words 'role-play', 'simulation', and 'gaming' have been used interchangeably in a lot of writing about classroom strategies. Although these activities overlap, they are not the same. The purpose of simulations is to try to copy roles and actions in a particular situation so that the mechanics of planning and decision-making will be understood. It is especially useful for learning about industrial decision-making and public enquiries: this may be the reason why most simulations do not run very well with our middle school pupils for whom these formal and managerial situations are not easy to understand. This will be discussed further in Chapter 6.

Role-play, by contrast, has more limited objectives. The pupils take on parts and act them out in order to feel their way into the likely reactions of the characters, as they perceive them. A 13-year-old boy is unlikely to make a convincing or useful job of acting out the part of a company director – he is much more able to throw himself into the role of a Victorian boy chimney-sweep. The historical exactness of the performance will be far from complete, but the pupil can often bring the character to life, extend his sympathy, and so begin to think about the causes of mistreatment, and means of curing the abuse. Thus role-play fulfils the essential aims of showing that science is about people.

Successful role-play requires information briefs for the preparation stage, but *not* complete scripts. Acting to prepared scripts allows for none of the personal empathic reaction that is the main purpose of role-play for STS.

Dr Jenner's vaccination of young James

The outline of this role-play is also to be found in *Exploring the Nature of Science* (see above). The following discussion does not repeat the information already published, but shows how the character briefs were designed so that teachers may use the general principles to design their own. It will also report what was learned during the extensive trialling of this unit.

The outline of the story of how Edward Jenner chose an 8-year-old boy and inoculated him first with cow-pox pus, and then with active pus from a smallpox victim is well known. When the resource material was being put together the television series *Scientific Eye* had just produced a 5-minute humorous cartoon of the story in their programme on 'Mini-beasts and Disease'. It contained visual jokes, such as cranking up the cow for milking, and was always received by the pupils with much hilarity, but it did not touch at all on the social or controversial implications of the story.

At first we used this cartoon video simply as a 'text' which was looking for a DART activity. We tried posters, but with only limited success. The pupils were attempting to copy the images they had just seen on the screen, and like all copying, it operated without further reflection on the subject matter. Taking a written text and putting into the pupils' own words, probably the commonest of all DARTs, similarly requires little involvement or creativity from the pupils. It is also poor learning experience. There seems to be a point here for quite general application – only if the pupil activity 'translates' the information into a new medium, will durable learning take place. (Acting the story, as it was seen on the television excerpt, was another DART used by one school but it also involved too much copying to make for sound learning.) The two most useful DARTs, in terms of the durability of the learning produced, turned out to be sequencing sentences about the story in terms of the scientific process involved, and a *Newsnight*-type role play. It was the second of these which had the strongest STS elements.

We have evidence from interviews carried out several months later with only average ability pupils that the details of this story were well remembered, and so were the scientific processes. That in itself is satisfactory. However STS teachers may want to take the general issue of testing new

Each group of pupils was given a brief for one of the seven characters who was going to take part in the final News Conference. The mother of James, for example, was told that Dr Jenner had failed to save two of her babies, or to help her sister who had died in child-birth. That would have been commonplace in the eighteenth century and might well have affected her trust in 'the good doctor', as the cartoon film had called him. James' father was told that he was Dr Jenner's gardener and lived on the estate in a cottage owned by Jenner. That in itself might have influenced him to allow the risky vaccination experiment on his son to go forward. The group that are to choose one of themselves to act the part of Jenner himself, read that he had already tried giving smallpox to a group of people who all said they had suffered from cow pox but two of them did contract smallpox and became dangerously ill. The other characters, seven in all, were 'created' to present different contemporary attitudes towards smallpox and its treatment.

The groups of pupils studied the briefs for 10–20 minutes and thought up answers to the simple questions set. This ensured that they had a little time to prepare for the flood of questions which might be thrown at them. At the last moment one person from each group was chosen, by the pupils or the teacher, to be the character in the press conference. The rest asked 'journalist'-type questions, in turn, from 'the floor'. In this way all the pupils got a chance to take part in the *Newsnight* Conference, which was sometimes chaired by the teacher. It has almost always been fun for all concerned.

medicines further and to talk with their class about how such tests are carried out today.

Pupils who had just taken part in this role-play were very ready to argue against the use of a child in any potentially dangerous test. We found it almost impossible to convince our pupils that children's lives were thought to be more expendable in a century when so many were expected to die in infancy. Indeed there was a consensus in one Year 7 class that the experiment should have been carried out on an old person who had less time to live. (This made us feel aged and uncomfortable!) The same criterion was applied in the subsequent discussion of the use of animals in the testing of new medicines. The pupils wrote that 'only old animals should be used'.

It is valuable to see young pupils offering ideas on science-based social issues in the classroom.

The discussion is of intrinsic worth because it attacks the damaging idea, which many pupils hold, that scientists, science teachers, and science itself, are callous and uncaring. At some later stage (see Chapters 5 and 6) the subject of testing drugs may become the focus of serious teaching for an STS course, for now it has just begun to build up a more humane image of science teaching.

CHAPTER 5

The examination classes

Citizenship at the top of the school

In the high theory of education, from the time of Matthew Arnold until the present day, preparation for responsible citizenship within a democracy has been the stated goal of all education (at least for the middle and upper classes of society!). And yet it has usually been only in the rather remote regions of the restricted-access sixth form that anything like this kind of education has been seriously attempted.

In most of today's comprehensive schools there is an abrupt break between the lower or middle school, and the sixth form. It is exclusively the sixth form which is allowed any measure of self-governance, permitted to choose their own forms of dress, or is addressed by the teaching staff with anything like the consideration expected by an adult of another adult. All of these are essential preparations for becoming an autonomous citizen, and are also signals of its approaching fulfilment. No doubt one reason for this sudden onset of emancipation is the voluntary aspect of post-16 education. Another reason may be the proximity of these students to the age of political franchise.

STS education is deeply impregnated with notions of citizenship and social justice. It is not surprising, therefore, to find it more strongly represented in the sixth-form curriculum than in any other parts of the school. It began there, in Britain during the 1970s, with courses such as Science in Society and SISCON-in-Schools and is now furnished with newer courses, largely derived from these, such as SATIS and Science Technology and Society both of which can now be taken at two different levels (GCSE and A level). These correspond, in levels of difficulty, to the two populations at the top of the school.

When the sixth form first showed signs of expanding, in the 1970s, by taking in more than just those preparing for university entrance or staying on for a term to re-take public examinations, several inspirational books about the 'New Sixth' were written. The best known of these, by A. D. C. Peterson (1973, pp. 30–31), put the educational implications of the new and expanded sixth form in the following terms:

> . . . every sixth-former, whether in school or out of it, needs and wants to develop his [sic] capacity for interpreting his environment, for understanding life. This means not only his external environment, the social and to a lesser extent the physical and technological environment in which he lives, but the inner environment of his own personality . . .

> But the young do not merely want and need to understand their environment, they want to operate within it, they want to be able, in some respects at least, *to be able to change it*. [emphasis added]

A later book by Reid and Filby (1982) traced the history of the sixth form from its origins in the public schools of the nineteenth century to the expanded sixth form which, by this time, already existed in the comprehensive schools. The sixth form was only just beginning to feel the push towards vocational training and work experience which figure so prominently in current post-16 education, but the less academic 'one year sixth' which had some vocational flavour, had already overtaken the traditional A level sixth form in terms of sheer numbers. The authors of this 'essay' were critical of many of the elitist syllabuses in existence, and finished their account with this comment (Reid and Filby, 1982, p. 246):

> Ultimately, the choices we make about sixth form education are choices about what form we would like democracy to take. Should it be one which leaves effective power in the hands of an elite marked off from the rest of society by a curriculum based on principles of exclusion? Or should it be one which sets a high value on the incorporation of as many of the population as possible into shared conceptions of democratic citizenship?

However much we may agree with that point of view it still offers no reason for the sudden change, described above, which marks admission to the sixth form in Britain. Education for democracy, if it has any virtue, should be a gradual process through which students are led by degrees, and by the teacher's encouragement and expectation, towards an understanding of how they can participate in democratic government. Much of this chapter will be concerned with the construction of syllabuses, and their examination, but the underlying theme will be education for citizenship. The arguments will override considerations of age and level within the school and may easily be adapted for use in sub-16 classes. Whether it is exclusively for the segregated sixth form or for students in any class who exhibit a more mature attitude towards social issues, the passages about sixth form education (quoted above) are important. They make two controversial points which are central for any consideration of the type of STS education required for our older students. Peterson writes of the students' need to be able to change society (see also Skilbeck's notion of education for 'societal reconstruction', 1984, p. 16). Reid and Filby question the separation of the two sections of the sixth form – the specialised academic group, and the growing numbers of less able generalists – in the context of their education for democratic citizenship.

For the academic sixth form

When STS first began in an organised way, this was the level at which it happened. Indeed the Science in Society course was not only intended for A level students, it was also piloted, for the most part, in the public schools. With the elitist tradition of such places there was a suspicion that it was being offered for the future 'captains of industry'. At least one of the founders of this innovative course spoke of his boys' interest as lying most naturally with the functions of industrial management. If power in our society was increasingly to be found in the boardrooms of the sun-rise technological firms, then it was possible to see STS as a preparation for the exercise of that power. This was the technocratic approach to STS.

That kind of STS education is rare in schools today. One reason for the shift away may be the movement of affluence in our society from industry to commerce, during the last decade. Another reason is, undoubtedly, our greater understanding of the real educational needs of the older student. Seen from a restricted access sixth-form college STS has much to offer which relates simply to the student's own development of study skills, to their understanding of academic work, and to their entrance into higher education. The following extract is taken from a report by a gifted and energetic sixth-form college teacher who has pioneered several new developments in STS.

STS for a closed-access sixth-form college

> As science teachers we have all had to deal with large numbers of students struggling with the intricacies of the Periodic Table, the structure of the atom, etc. with little motivation and even less clue to the relevance of their (sometimes!) valiant efforts. This may be less of a problem to the able who can take a more abstract and analytical view of the subject matter when it is presented in the traditional format. What is needed for a lot of students is a fresh approach and one which catches their imagination. STS is one of those ways.
>
> The students, whether very able or just average, have many of the same pressures. During the two years of sixth-form education they mature at an

accelerated rate. Not only are they following academic syllabuses but they also have part-time jobs, more responsibility at home and more chance to make many decisions for themselves. At college too, demands on them have changed: they are asked to give their own views on topics, to access data for themselves, to assess their own academic progress, to evaluate courses, and to think about their futures in further education or in the world of work. They are expected to develop skills of communication, to learn to deal in an adult fashion with peers and staff, and to integrate their increasingly active social lives with their work. Needless to say, students at sixth-form level are, these days, much less likely to accept courses which are not relevant, enlightening, or enjoyable.

The role of STS in a closed-access sixth-form college is twofold:

(a) It provides a one-year GCSE course for those students who wish to study only two A level subjects. The course allows them to take a science qualification at GCSE if they do not already have one. For those who do have a GCSE science qualification the STS course allows them to study the social aspects of science which, although increasingly touched upon at GCSE, may not have aroused their interest previously.

(b) It provides an AS [Advanced Supplementary] course for those students who are opting for the [two A level + two AS level] route to Higher Education. On the whole these students have only one subject which has any science flavour, such as Social Biology or Psychology. Those who are taking two A levels in the traditional sciences are more usually advised, although not by me, to take an AS level in Mathematics or, perhaps in a modern language to further their business aspirations.

I believe there is an enormous scope for STS at AS level, but many sixth-form colleges are yet to be persuaded of this. They see AS level courses as ones which replace traditional A levels. But STS is not that sort of commodity. It is 'an extra', not a replacement, an additional subject which broadens and complements sixth-form science studies in a new and essential direction.

Although the target groups are different, both these STS courses fulfil many of the requirements

for entry to university and polytechnic. The challenge is to overcome the fears of college teachers and to get the subject more widely established. There is clearly an interest at the admissions end. Students are coming back from interviews having spent a good part of the time discussing STS. Presumably the interviewer wants to know more about it!

There is no typical lesson for STS. Some topics lead to discussion in greater depth than others and the teacher will learn to 'go with the flow', after having created the stimulus to learn. It is essential to have an interactive dialogue of views at this level to dispel the myth that 'teacher knows it all'. There should be an equality of access to information so as not to create a power imbalance, if students are to develop their own views and decisions. If everyone is contributing to the discussion and analysis of a problem, then the power-base is more equally distributed.

These sixth formers are often unaware of the effects of science and technology, and of the decisions that scientists and others have to make. Without such awareness they cannot challenge established views. They often start with a low sense of their own power. They have their own private and often untested assumptions about the way society is – about full employment, and about health care in the developing countries. In STS courses these, and other assumptions, are all likely to be challenged.

It is important to allow for flexibility. The teacher must be free to present material in a manner suitable for her own students. The following are inherent in all STS teaching:

● The subject matter is relevant to the students and to the society in which they live. (This I believe to be the key to success in all science teaching.)
● It creates new interest in science and technology.
● It offers a preparation for living in our increasingly technological society where the less well informed can easily become disempowered.
● Above all else it gets the students thinking, discussing, and *questioning, questioning* . . .

STS for our ethnic minorities

The next extract has been contributed by an STS

teacher who works in a girls' school with an over 90 per cent Asian intake from a working-class inner-city district.

Our girls, it seems to me, have less awareness of the world they live in than do the majority. This is not surprising. Their parents are mostly working class and do not speak much English and this makes it difficult for a lot of the girls to grow up feeling that they are really a part of the British society. They have sometimes said, about issues they have discussed in STS lessons, 'Isn't it a good thing we do these lessons otherwise we would never hear about this!'

I have noticed an improvement this year with my STS group. They are now the brighter ones who are doing one or two A level subjects, often in Sociology or History as well as STS. But even so I have often found myself delivering knowledge that other young people might well get from their homes. In science lessons the girls perform well, but they have simple and unsophisticated views about social issues. At times it can be almost a battle to start them thinking for themselves at different levels about problems. They have opinions, as everyone has, but I doubt if they have worked them out for themselves because, I be-lieve, they would find it very hard to disagree with what they are hearing at home. At school, in their STS lessons, I can see this process of forming their own opinions gradually beginning.

It is especially hard to teach them about issues which are really controversial – but I enjoy trying. Sometimes I show them a serious video from television; it might be about health-care, vacci-nation and the resulting increase in population in the developing world. Of course that is just one little part of the equation: there are other factors to consider like culture and tradition. During class discussions I may have to act as 'Devil's advocate', but I do worry that I may over-influence them. At present I have a large group with a wide ability range for whom special resources have been pro-vided so that they can work independently at their own pace and level. But I am finding that even this makes some of my girls feel insecure. (They would probably rather that I dictated pages of notes!) One strategy I have developed is to give them a series of questions which they discuss, and prepare a report. Then I tell them to consider the same questions and prepare a new report from the

opposite point of view – totally biased. They are not good at balanced arguments and their first attempts have almost always been biased in the other direction, so I am beginning the process of examining the assumptions inherent in their own arguments. I set the tone of the whole course in this way. The girls need to get used to considering both sides of any argument.

I find that they almost always choose to do topics like health and food in the developing countries. They probably relate more easily to that because of their own Asian background, and they actually find it harder to relate to things which are happen-ing around them here in Britain. Their under-standing of our technology, for example, is rather limited. Their community has a very strong work ethic, and for them technology is often equated with unemployment among their menfolk. Many of my girls will never themselves have the oppor-tunity to work outside the home. I do not want to impose my culture on them but I find it frustrating sometimes when I see that they are not readily given the means to make any choices, even at home. It is not due to any lack of intelligence on their part, but because of a whole range of con-straints, what they learn in science at school does not easily become a part of their lives, or their thinking, at home.

Technology and decision-making

Their narrow view of technology – just computers and electronics, or automation and unemployment – may be partly due to the fact that most of them have never played with construction toys when they were little, nor been allowed to get dirty at play. Often women do not even do the cooking at home. So I have sometimes set a 'survival exercise' on paper to begin to give them a new perspective on thinking and doing. I set up an imaginary situation, like being ship-wrecked on a desert island, and provide a list of items which they find and can use to build a shelter. This is designed to show technology as a way of dealing with basic problems using local materials. They may not do it very well at first: most of the girls tend not to be very confident with their hands even though many do quite wonderful needlework at home. But they do make progress.

Above all I want them to come to understand that they can make their own decisions and choices about a whole range of issues. They can even do this at home. Take the Greenhouse Effect, for example. They come up with the usual talk about damaging the atmosphere and that we shouldn't cause this pollution. When we discuss it together in class I get them to tell me about the small things which they could do themselves, in their everyday life at home, which would contribute to protecting the atmosphere. This becomes their own way of making a contribution towards dealing with a very large problem.

Extending horizons

The girls all tend to be very suspicious of politics and politicians. Basically they believe that all politicians tell lies. That's not good enough. They have got to learn to sift through what they hear, to think about it and make choices. For this they need to start reading quality newspapers and listening to more serious programmes on the television. They need to give up depending on me, and learn to ask questions for themselves.

For our girls I have always thought that STS was one of the most useful things that they have done in school. When you think about their special problems with language and literacy, it is surprising what a lot they get out of the course. I mark their project work and its standard is often impressive. They sit an examination which, for them, is really quite hard; but they work, and they achieve really quite good results. It is just because their lives are so sheltered at home that an STS course becomes so immensely valuable.

As if to echo these sentiments an Asian student wrote to her teacher after the end of her year of STS:

As an Asian person, doing the STS course showed me that this [being Asian] did not alter how I approached things. And if I really wanted to make a difference what I could do. It encourages me to be more active in the issues that we have studied – trying to save the rain forests, and raising money for health-care in the Third World.

An examination syllabus in STS

These two extended comments, by experienced STS teachers preparing students from such differ-

ent backgrounds for public examinations, set the scene for examining the examinations. Here we need to set out our objectives so clearly that a syllabus can be constructed, papers written, marks allocated to candidates' scripts, and standards maintained from one year to the next. This is altogether a more rigorous task than was attempted in earlier chapters and will be tackled in four stages.

The usual method is to begin by setting out the *general educational aims for STS* – 'pious aspirations' as some have called them rather dismissively. They may only be forgiven for sounding too supercilious to be examinable, if they set the task in a useful context. They represent our *purposes for teaching STS*, as a directive for those who begin to design the syllabus. Why STS? The list of general aims are:

- To increase citizen's scientific literacy.
- To help students become better decision-makers.
- To encourage interest in the interactions between science, technology and society.

Each one of these is to be found in one or other of the current published STS syllabuses. At first sight they seem to provide little beyond the kind of 'motherhood' mottoes which sound more like maxims than educational guides. Nevertheless, they are not quite without use: it is quite possible to use such statements to evaluate the next set of aims which are far closer to classroom reality.

For this type of specific aim we might use the five *areas of interest* which were set out in Chapter 1 as follows:

1 Environmental threats, including global ones, to the quality of living.
2 The economic and industrial aspects of technology.
3 The fallible nature of scientific explanation.
4 Personal values and group concerns about the uses of technology, leading to appropriate democratic action.
5 The multicultural dimensions of technology.

These are much nearer to providing a syllabus – we can begin to see what students might be taught.

The first set of general aims can be used as a checklist for these more specific aims:

AIM 1: 'Scientific literacy' may be principally delivered by (3) in the list of areas of interest.

AIM 2: 'Better decision-making' will call, at the very least, for (1), (4) and (5).

AIM 3: 'Interest in the interactions' will be encouraged more, perhaps by (1) and (2).

Every item on the list has now been accounted for in terms of the list of general aims. However this correspondence cannot guarantee that the trawl of objectives has been sufficiently wide, only that the five specified are all useful.

Another way of setting out the areas of interest is to start from the three areas of STS – science, technology and society – and decide what students should know about each. This is something like what has been attempted by the SATIS 16–19 'Framework' or F-units (SATIS, 1991).

The science in STS
     What are scientific theories?
     Where do these theories come from?
     Society comes to depend on its theories
     How scientific theories change
     Science and para-science
     How scientists make scientific knowledge
The technology in STS
     What does technology mean?
     Inspiration, invention and science
     Technology in industry
     Technology and economics
     Cultural differences in technology
The social decision-making in STS
     Risk perception
     Controls and regulations
     The process of government
     Public groups in decision-making
     Individual understanding and decision

These are more detailed than the former five areas of interest, which makes them useful for designing student textbooks.

A subject syllabus needs even greater precision. We shall need to think about *assessment objectives*. These are the clear indications of what students could be expected to come to know and understand, and skills that they may be expected to develop. To list these components we shall need both to look backwards at the general aims or purposes and the areas of interest, and also forwards towards setting the examination paper and marking the candidates' scripts. Such requirements are hard task masters. They mean that each assessment objective will need to be judged on at least three criteria:

● Does it fulfil some part of our purpose in teaching STS?
● What precise knowledge or skill is involved?
● Is it both teachable and assessable?

Most lists of objectives contain *knowledge items* such as:

● Some key scientific concepts
● Structures of civic and industrial organisation
● History of science and technology in some period
● Some terms used in logic, and in technological discussion

Each of these can then be defined more closely and placed in the actual syllabus (e.g. which concepts? how much about industry? which period of history and what detail? should we include specific philosophical theories?).

Next comes a list of *skills of analysis*:

● Be able to understand and interpret statistical data
● Be able to read and interpret articles
● Be able to relate data and information to a problem

This list is also useful and can be translated almost immediately into teaching strategies. It will guide those who set the examination questions as well as those who teach the course.

Some syllabuses also try to specify *skills of evaluation*. Indeed it is hard to see how the purposes of STS, which set such store by helping students to become better decision-makers, could be translated into a syllabus without including this more diffuse and intangible area. The STS syllabus produced by the Northern Examination Association (1990) for example, requires that candidates should be able to

- weigh evidence and to identify inherent assumptions including value judgements,
- propose alternative ways and means of solving problems,
- assess the costs and benefits of alternative solutions including possible social consequences, and give reasons for a choice of one or more solutions.

The difficulty with these objectives lies not in bringing them out into the open in our teaching (see, for example, the section above by the teacher of Asian girls), but in examining for them. The first evaluation skill above simply asks for the identification of hidden assumptions. It is important that these are not denigrated as 'non-scientific' or not 'objective' as they are in some other syllabuses. *All evidence will incorporate more or less hidden assumptions. None can be value-free.* Our students should get used to looking for assumptions and evaluating them through the use of their own value system. Some teaching materials, and even examination questions, have deliberately included passages where the author's intentions can be rather easily exposed and discussed. This is not done in a spirit of 'the true scientist must be objective', or even the relativist's dismissive 'all value judgements are equally permissible'. It is an opportunity for reflection on moral standpoints, presented as an integral part of the STS lesson, and a challenge to evaluate one's own position.

The second evaluation skill presents another problem. Many of the STS issues which are here called 'problems' will have already taxed the best brains in the nation. Could the students' 'alternative solutions' possibly be of any practical value? It is important to notice that this NEA syllabus does not specify that the students' solutions will resolve the problems, but they do need to be sensible. (The development of the third of the evaluation skills would imply that each suggested solution has been subjected to some fairly careful scrutiny by the student.)

The third evaluation skill has been more helpfully worded, even though the almost insuperable

difficulty of identifying all the social consequences of solutions to our technological problems is precisely what has led to so many modern dilemmas. Fortunately the last part of this requirement gives both teacher and examiner much clearer guidance. *Students should be encouraged to give reasoned choices.* The reasons they give may well be sound and worthy of good marks, even if the proposed solution does not seem entirely satisfactory either to the teacher or the examiner.

Even after guidance has been given on the written paper there is still unfinished business. Some of the general purposes of STS cannot easily be satisfied by the traditional format of examination. Is it really possible to

- encourage interest in the interactions between science, technology and society, or
- express personal values and group concerns about the uses of technology, leading to appropriate democratic action?

Many examination boards have decided that it is only through student's own project work on a freely chosen topic that the first of these aims can be met. This is now a popular addition to almost all GCSE assessments. Some have suggested that small group discussion work would be the only way to satisfy the second of these aims. (This is a far more innovative practice which will be discussed in greater depth in Chapter 7.)

Feedback from STS examination scripts

Students' work in examinations provides limited amounts of evidence. In the first place there are the official reports from the Chief Examiners; in the second a few very limited but more detailed explorations of the success of candidates judged by their own scripts. Only answers to that part of the paper which examines the candidate's general understanding of STS will be scrutinized. The work on specific options within the paper is too factual to give answers about the real difficulties or successes.

Examiners' reports regularly (a) praise the candidate's efforts to understand and interpret comprehension passages, (b) bemoan the fact that their interpretation of graphical or numerical data is of a much lower standard, and (c) comment on the difficulties candidates seem to have in answering questions about the *nature of science*. Of course this is not a blanket effect. It is often reported that questions on Darwin's theory of evolution (option on 'Evolution, Genetics and Fertility') are well answered, whereas the same candidates often have greater difficulty in answering questions about scientific theories in general.

In a similar vein, examiners report that questions about health-care in the candidate's own country are often answered with both care and understanding, while questions about health-care in the developing countries are, in general, poorly answered because they are based on prejudices about the inhabitants' ignorance, lack of education and malnutrition. This can even distort the candidate's interpretation of facts clearly presented in passages on the examination paper. (This problem of the reception of data being coloured by expectation is explored in Chapter 7.)

Examiners also report on the project work done by the candidates, often praising its high quality and evidence of commitment. They make three general points for the guidance of teachers.

1 Many students need help in choosing an appropriate topic. Some, like 'Alternative Energy' are too wide; so is 'Health in the Third World'. What is needed is a subject which interests the student but is sufficiently restricted in scope for her/him to become, as it were, a 'mini-expert'. Then there is some possibility of students achieving adequate coverage – not so wide, but deeper.

2 The topic chosen should allow for aspects of science and technology and societal effect. Detailed description of some new item of technical progress, such as a new car engine, may be too technological. Discussion of delinquency may allow for too little in the way of science or social effect. Topics such as 'Wind-power' and, more surprisingly, 'Death, burial and cremation', have managed to produce well balanced topic work.

3 Comment and personal evaluation are essential. Too often all that is offered is a brief summing up of points of view 'for' and 'against' on the last half page. The project becomes far more balanced and readable if the various social perspectives are presented earlier. This makes for greater ease in suggesting alternative strategies.

Research data

A question on a paper for 16-year-olds, asked them to 'describe any one school experiment which helped you to understand a scientific theory'. Most of the students found little difficulty in describing an experiment, but identifying any theory at all – let alone one associated with experiment – defeated most of them.

● 25 per cent correctly identified a theory related to the experiment
● 27 per cent gave the experimental result as if it was a theory (e.g. 'that copper conducts electricity' or 'the Theory of Brownian Movement')
● 24 per cent gave the experimental process – the food test or the construction of a solar panel – as if it were a theory.
● 25 per cent omitted the question altogether even though it was in the compulsory section.

From these sorts of results, which can be matched in the 'Nature of Science' literature (e.g. Aikenhead *et al.*, 1987), it must be assumed that the connection of theory with experiment is very rarely taught. Only for the specially contentious topic of evolution is there evidence that science teachers are making efforts to show how scientific theories are constructed. However, this is a somewhat special case, and may offer students little of more general value. 'Experiments' in evolution are rare and difficult (for one good example see *Hen's Teeth and Horse's Toes* by S. J. Gould). The creationism debate, which is entirely relevant to STS courses, is easily highjacked to be about the nature of truth, or science versus religion, instead of the uncertainty of evidence and tentativeness of theory, which it illustrates so very well.

Research findings also indicate that there is

general confusion about the meaning of 'technology' (see also Breakwell *et al.*, 1987). As the passage quoted earlier from the teacher of Asian girls indicated, many of our young people simply equate technology with computers or electronic machines. In Solomon (1988) the examination candidates were given the following question:

> . . . the new technology of *in vitro* fertilisation outside the human body is a response to the social problem of infertility. However the development of this technology has resulted in problems concerning the use of live embryos in scientific experiments.
>
> (a) The paragraph above uses the term 'technology'. Explain carefully what technology is.

With help from the preamble, the question a satisfactory 57 per cent of the students managed to avoid the usual pitfall, and wrote instead that technology was 'a process using knowledge for a social purpose' (in some equivalent form of words). Even with this help, 16 per cent of the candidates still defined technology as equipment, tools or machines.

The second part of the question asked:

> Give one argument for, and one against, banning experiments on human embryos. Give you own view in relation to these arguments.

The fact that 85 per cent of the sample of 284 managed to construct plausible arguments from opposing points of view is impressive evidence of the careful teaching that had taken place, and nicely validates the opinions and experiences of the two practising teachers quoted at the start of this chapter.

The research article ends with comment on how few of the candidates, only 12 per cent managed to suggest ways in which new laws or regulations might answer the problems and anxieties which the technology had aroused. This brings our attention back to the citizenship aspect of education which is the principal educational purpose for all sixth-form curricula.

There is just a little more data on this important point which is taken from a more taxing paper on Science, Technology and Society designed for more able students taking the AS paper. Once again the candidates were asked about scientific theory.

> It is not always easy to tell the difference between the following kinds of statement:
>
> a scientific theory,
> a simple generalisation,
> a non-scientific theory.
>
> Give *one* example of each of the three kinds of statement, and explain carefully why you have selected it to represent that sort of statement . . .

This time 57 per cent of the students were able to give an example of a theory. There was some additional internal evidence to show that the ability to answer correctly may have been produced by specialised STS teaching. Very few candidates (less than 7 per cent) both got half marks or more on the question as a whole, and yet failed to give an example of theory. This may suggest that normal science teaching within school, of which these students had all experienced some five years or more, had not addressed the subject of scientific theory at all. Without such teaching students can only fall back on folklore about scientific evidence and proof, e.g. 'Scientific theory is a statement supported by scientific evidence. Example – that a new gas field exists in the North Sea, for which there is evidence'. What that lamentable answer shows is how necessary examples of theory are for showing the student's understanding of terms like 'evidence'. It is sometimes suggested – usually by those who have not tried teaching STS – that the social element is particularly easy and needs no teaching. Since it is here that the subject comes closest to studies in citizenship science, it is valuable to explore the responses to a question of this kind.

> . . . There have been several environmental issues in the last few years which have raised a great deal of strong feeling. Sometimes scientific 'experts' have even disagreed about the 'facts' of the case.
>
> (i) In the case of one such environmental issue explain why the experts might have disagreed.
> (ii) If members of the public want to take part in the controversy over the issue, what possible course of action can they usefully take?

(iii) People often blame the media for stirring up controversy. What is your view on this?

These questions did not prove to be so easy. In the first part, 30 per cent of the students wrote simply that the experts would be biased. Some just suggested that experts might be corrupted by the industry for which they worked or by the money that they hoped to gain. Just 44 per cent of the answers indicated clearly that the evidence was, by its very nature, unlikely to be conclusive. In this way teaching about the nature of scientific knowledge was being vindicated. It supplied essential understanding of its controversial nature.

In the third part of the question, 35 per cent of the answers simply agreed that the media was 'sensationalist' and 'stirred up the muck'. (Newspapers were blamed for this even more often than television suggesting that these candidates were not reading the better quality press.) This paper was being sat at a time when this pejorative view of television coverage of news was being widely discussed. It was encouraging, therefore, to find that 56 per cent of the answers both agreed that the media brought issues out into the open, and also added their own comment that this was valuable and even necessary for the public.

In conclusion, it is worth returning to the accounts given by the two experienced STS teachers who described how they taught their students. Both reported how difficult it was to get them to question what they heard, viewed and read; how important it was to get them to examine both sides of public issues and not just to allude vaguely to 'bias'. It seems, from the limited examination evidence, that they were on exactly the right educational track.

CHAPTER 6

Games, simulation, and role-play

Industrial awareness and value issues

There are at least two aspects of STS education which have been given scant attention so far in this book. One is concerned with economic and industrial awareness, and another is about the students' discussion of values with respect to STS issues and possible democratic action. These two aspects of STS are miles apart, with one set in the hard-nosed world of industry and commerce which is largely unknown territory to the students, and the other having its existence so deeply within their own feelings about what is right and wrong, that articulation may be hard for quite different reasons, related to shyness in talking about feelings, or lack of appropriate vocabulary. What binds values and industrial awareness together is only that in STS issues such as pollution or power generation, both are involved. In addition there is a view, which will be challenged in parts of this chapter, that learning about both – industrial management and personal values clarification – can be handled in the classroom by the same strategies of role-play and simulation.

There are those who deny the need for any special technological or industrial understanding in dealing with science-based social issues. Mary McConnell (1982, p. 13), for example put her view in uncompromising terms:

> The problems associated with technological development may not primarily be problems of technology. Rather, they may be problems among us and between us, problems in a large measure the

result of a pluralistic society in which common purpose, common values, and common images are no longer present. They are issues involving conflicts between values and goals within persons and among persons, rather than conflicts between dams and people, or industry and the environment. Resolution of conflict requires communication and creative problem solving to facilitate mutual understanding and effective interaction between people and groups that have different values, different images of the future, and different images of trade-offs and benefits and costs.

We may all agree that in our pluralistic society there are great potentialities for inter-group conflicts of interest and values, but might not be so sure that the basis for resolution of these is simply to be found through communication about values. There is a suggestion here that students do not need to learn about technology, or industry, or economics. All they require is a nose for sniffing out trade-offs, benefits and costs which do not conform to their values, and a facility for effective communication. Sadly there is now a real conflict between industry and the environment in a sense which transcends the caricature of the evil polluter who does not care about people, and the spotless environmental purist. As people, the groups cannot fail to share many objectives; it could be lack of a shared understanding which separates them.

STS education is dedicated to the proposition that it is well worth learning about the knowledge and perspectives of the different groups involved in the technological issues of our times, for only with such understandings can decisions about the issues be made. We need to learn about the perspectives of those in industrial management, if at all possible, and how their world view is constructed. Just as we try to teach about the nature of science, so we should also strive to teach about the nature of industrial technology.

Games and imagined simulations

Ever since STS first broke in upon the school science curriculum in the late 1970s there have been calls to develop new and appropriate strategies for teaching. Several of the suggestions produced have involved gaming, and simulation. According to Ellington *et al.* (1981) exercises of this kind can fulfil objectives:

> . . . *educating through science* – fostering interpersonal and communication skills, and
> . . . *teaching about the nature of science and technology* – illustrating the making of political, social and economic decisions.

There is little doubt that the first of these objectives can be met by group discussions about industry or other matters, in science or in any other school subject; it will be considered in some detail in later sections. But the second objective becomes problematic if the political, social and economic decisions lie too far beyond the teacher's and the student's own experience.

A number of fairly light-hearted simulations of other countries' technological problems are available, which may deserve the name 'game' not just because they try to amuse students, but because their names and fact-sheets immediately show them to be thoroughly fictitious. Although much may be learned through fiction it is doubtful if STS, whose very basis is relevance to real social problems, can afford to use it. Early games such as *Minerals in Buenafortuna* (1983) used just such fact-sheets and encouraged students to discuss the issues that had been invented in rather contrived and restricted situations. Students were working with circumscribed sets of information and soon found that the agenda was already set by the

authors. Questions such as whether the country really wanted to develop its mineral resources were taken for granted. Although the authors stated that open-ended questions such as 'the consequences in terms of pollution, noise and the disruption of the local community' can be put to the group, the 'only problem of any importance', they admitted, was predetermined by the game-makers.

The word 'game' is bound to be misleading in a context where a simulation of reality is intended. It is possible to buy many of these, ranging from simple snakes-and-ladders to complex role-plays about newspaper reporting of a health hazard in the Third World. In the early 1980s there was even a dice and card game about subsistence farming published by Oxfam, in which participants picked up cards announcing that they were suffering from severe malnutrition, life-threatening diseases, or even death. It was called, with quite startling ineptitude, *The Poverty Game*! It has now been re-named *The Farming Game*. Although it almost always caused hilarity in totally inappropriate places, this game did serve one purpose for which it would be hard to find its equal. When the players reached a stage at which their village was so deeply sunk into poverty that at long last they were permitted, by the rules of the game, to pick up a 'Help' card from the pack, there were occasional rebuffs when the legend on the back stated starkly 'Western governments do not like the politics of your country – NO AID'. The students' disappointment and muttered comments of 'not fair' made a brief but instructive point.

What these imaginary scenarios were really attempting was (a) a simulation of a *nearly* real situation, and (b) a role-play by students. The students were asked to act out particular roles in order 'to appreciate hard decisions made by others' (Ellington *et al.*). This created three serious problems:

1 The first was that most students would be making their own difficult future decisions as citizens rather than as technocrats so the activity missed that immediate relevance for which STS education aims.

2 Secondly, in so far as the simulation called for role-play of characters whose background knowledge and skills were largely unknown to the students, in practice it produced enormous difficulties.

3 The third problem was that acted parts call for artificial opinions. Participants who already had formed clear, and even passionate, ideas on the public issue in question, might be precluded by their given role from expressing what they really believed.

Much has been learned about simulations and other classroom activities since those comparatively early attempts. Nevertheless some of the problems of designing and using simulations still remain, most noticeably those relating to industrial understanding.

Teaching about technology, economics and industry

Economics is a popular subject in the sixth form with ever-increasing enrolment. It also has a part to play, at this level, in STS courses but not as a dry clear-cut mathematical or theoretical discipline with complex modelling and a right answer which relates more to industrial growth than to quality of life. Nor is it of very great value just to burden all the students with endless graphs of imports and exports related to some technological innovation. We shall, of course, want all our students to be numerically literate to the extent of being able to scan information from graphs and charts of various kinds, with ease. However it is the use to which such information is put which distinguishes STS from economics for its own sake.

Economics, like science, is just one part of the social dimensions of technology with which our courses will be concerned. It is a weighty factor for industrial bosses to take into account when they make decisions. Although teaching about technology within industry has been considered here, alongside simulations and role-play, it is by no means certain that these are the only ways, or indeed the best ways in which it can be taught.

Stories of past innovations are an excellent way to illustrate important points as well as being very inspiring to some types of student (e.g. Goodyear and the vulcanisation of rubber, or Russell Marker's explorations for making an oral contraceptive). From these stories, so long as they do not amount to any more than heroic legends, the following general points should emerge:

- Scientific knowledge is not always essential for invention, but it will accelerate its later stages.
- Chance plays a part in invention, but events still have to be noticed and exploited by the inventor.
- War often triggers and accelerates innovation.
- Discoveries may be made by several inventors at the same time because of market drive.
- New technologies change people's ways of living in both expected and unexpected ways.

Other aspects of technology also need teaching. We shall want our students to learn about cultural differences in the technologies appropriate for different countries, and the economic constraints which operate there. Closer to home they will need to learn a little about the role of technology in modern industry, which is substantially different from the tales of individual inventors mentioned in the previous paragraph. This will involve some practical economics related to the costs of R & D (Research & Development), competition after the first phase of innovation which is protected by patent, and military spin-off into civil use. It is here that simulations of what it might be like to make decisions within industry may be able to breathe some life into these rather dry and esoteric notions.

Industrial simulations

Most teachers find industrial simulations extraordinarily difficult to run in class, not just because they themselves have little grasp of economics, but because they have so little feel for the human context which they are trying to simulate. This becomes most apparent when they need to show their classes what kinds of argument would win the day when rival projects jostle for funding within an industrial firm. Some teachers have found an excellent local solution to their problems.

One very experienced STS teacher told me how she had once set up a hospital scenario where she and her students could study the decision-making concerned with priorities for operations such as kidney transplants and hip replacements. She had selected the problem from some resource materials. The students, using costings from the printed example, had already spent about a week trying to decide what treatments and operations they would put money into. Then she invited someone from the management team in the local NHS hospital to come in. Given the same problem she asked him to tell her class how he would have allocated the money.

> 'And, do you know', she said 'it was *totally* different from anything that they or I had come up with!'

But when the manager made his presentation and went through his reasons it had all seemed completely valid to both teacher and students. They could immediately see why he had made his decision.

> It was fascinating. I could not have done that sort of exercise and given that kind of reply myself. Economic and industrial awareness are important because they influence decisions which directly affect society. I get the impression that decisions are made as to whether or not a particular product will have R & D money put into it for reasons I would never have imagined. It goes ahead for important internal reasons, not just as a result of simple market research.
>
> I have never found role-play very successful in my classes unless I've already got some familiarity with the background. Role-play in industry is particularly difficult. For GCSE courses I can make do with a computer program showing, for example, the costs of metal extraction, or the advantages of one site over another. But for more advanced work you do need to have a proper science and technological link. I think that most areas have them now. In my area the 'neighbourhood engineer' is from the polymer industry, and he has always been very keen to come in and talk, not about the science involved because I can do that, but about the decision-making processes within industry – and that's exactly what I need.

Other STS teachers and resources designers have held similar views on the need for knowledge about industrial decision-making. New course materials (e.g. SATIS 16–19) are appearing for the sixth-form age group but it seems possible to discern the same kinds of problems in most of the units which aim at industrial simulation.

One simulation (Unit 4), 'R & D at MUPCorp' (from SATIS 16–19), is said to explore 'the various decisions which must be taken after the initial scientific discovery of a product or process, up to the point at which it might become a commercial product'. This seems ambitious but it leads to a short and well-devised activity. The students take up roles as Project Managers with the task of deciding which of a number of 'promising' research projects should be explored further with a view to future investment. For each project they have to consider at the very least:

- whether it will work on a larger scale
- what uses it may have
- how big the market might be
- how it fits in with the firm's existing expertise
- social effects
- scale of manufacturing plant
- promotional ideas for advertising

The printed list is much longer and the tasks seem to be demanding.

There are ten new scientific research projects to be considered; all of them are comprehensible, at least in general terms, to a lay person, and most are thought-provoking and even amusing. The activity was written by a science teacher, vetted by some industrial scientists, and shows every sign of being fun to do in the classroom, with groups of students extolling the virtues of their particular product with the 'added spice' of a threat of redundancy if they fail to convince the Board!

The author of that activity adds a note from his own experience, and those of others, that 'it is important that students should not get "hung up" on the nature of the product. Scientific licence is allowed in advocating the merits of proposals which seem outlandish'. This point is thought-provoking. The purpose of this simulation, and others like it, is to experience some of the problems of industrial decision-making from an internal standpoint. From what has been discussed earlier it is clear that the gaming aspect of the activity could easily outweigh the relevant industrial aspects. Here again the presence of a 'neighbourhood engineer' might well be very valuable in turning the balance towards reality.

Another simulation (Unit 56), 'Planning a new edible-fats factory' illustrates an alternative approach, which would seem to ensure validity, by inviting industrialists to write the simulation themselves. (Several of this kind of exercise already exist but, in the experience of some teachers have not been easy to use.) This example, which runs to many pages of close script, is based on a real case study which is used in industry as a training exercise. Once again the students work in groups but, instead of taking on roles of scientific entrepreneurs, they have to become 'management consultants' who make detailed presentations about the best site for a new factory making edible fats, or 'specialist subcontractors' who give, or sell, specialist advice to the contractors.

This exercise has not only traded fun for reality and made a convincing simulation of a real situation, it has also asked the students to adopt specialised professional roles. Now the printed suggestions for running the simulation specifically include inviting a 'visitor from industry' to take part. This is clearly an essential ingredient if there is to be any chance of making such a difficult activity come to life in a school classroom. Teachers who have succeeded with this sort of simulation have often involved the whole of the sixth form for a complete day of 'industrial awareness' activity.

From inside industry or from outside?

Even if all the practical difficulties are overcome the simulation of industrial decision-making may still miss most of the central concerns of STS. (One teacher even commented that, in the previous industrial simulation, the basic STS question – 'Do we really *need* another factory?' – could only be treated after the activity is over.) The aims of STS

courses nowhere include the training of future managers for industry; but they have been reiterated often enough in this book for key terms such as 'citizenship', 'values' and 'democratic action', to be thoroughly familiar. Although all our students do need some understanding of the economics of innovation, most teachers in schools with a mixed intake will want to spend more time considering the point of view of the workers or the general public. Here it will be the student's values which suggest priorities, and not the rules and roles of the simulation.

This externalist approach would teach from the perspective of active and concerned citizenship. The students would be learning, for example, not only how to test for polluting substances in the environment, but also to value moral and civic reactions to the act of pollution. The following extract comes from the experience of an enthusiastic STS teacher from a working class area in London, who is talking about the work he does on industrial hazards with Year 11 pupils. The strategy, of course, can be neither simulation nor role-play if the outcome is to concern real social action. Instead he weaves STS into his school chemistry coursework, teaching about industrial controls in this country, and then contrasting it with what happens abroad.

We do a lot about hazards and the transportation of dangerous chemicals – how to deal with spillages, and controls in the chemical industry. We were talking about hazard signs and the kids were working out how to recognise the dangerous chemicals, and it seemed that everything was wonderful – the signs were clear and the fire-service well organised.

Then I said 'How would you feel if there was a chemical spillage and ten thousand people died?' You know, they just laughed, and said, 'No. It couldn't happen.' Of course over here it couldn't happen, but I was leading up to talking about the tragedy at Bhopal, and conditions in the chemical industries in the developing countries. This really took off. We have a lot of Asian children here at this school, boys and girls, and they knew nothing about it. They listened. All these people still dying in Bhopal. You should have heard how they listened!

It would be difficult to deal with issues as emotive as this through simulation and role-play. The aims would be completely different to those in industrial simulations which study the internal processes of decision-making. Now the students would need to have freedom in constructing the roles that they *want* to play, since the objective would no longer be to 'feel your way into someone else's shoes'. The new objective will be to work out what their own sense of justice and compassion demands in the kinds of circumstances where ordinary people may confront risk or disaster. The rationale for setting this kind of teaching in a simulation is that there is something in the procedures being simulated which could be especially valuable to the students as citizens.

Teaching about social decision-making

The final 'S', in STS, is too often assumed to be no more than a backcloth for the science and technology being taught. Curriculum developers justify their materials by pointing out how useful the innovation they are teaching about has been to society, how it has added to the quality of living, and increased industrial production. That, for them, comprises the whole social context.

Other courses develop more realistic resources which point out both the benefits and the risks in technological innovation. But it is precisely here that the missing ingredient becomes most acute. If there is an uncertain balance between risk and benefit, if the results will affect large sections of our society, and if there is controversy about the issue, we can hardly end our teaching with no more than a limp question mark as to the future. The young people in front of us need answers to the question, how can 'ordinary' citizens like them find their own voice in the debate? It is here that the final S begins to impose its curriculum.

The kinds of topics that need to be covered might run as follows:

1 Risk analysis including some simple calculations of probability, but also personal perceptions and toleration of risks, the considerations of special groups such as workers, and our obligations to native peoples, to wildlife, and to future generations.

2 Controls and regulations which are feasible, the 'teeth' which governments might give to those who enforce the controls, the powers given to public inquiries, and the problems with getting international agreement in the face of global issues.

3 The process of government – the powers of MPs, ministers, parliamentary committees and local councillors – and how they tackle matters related to science and technology. The relationship between the citizen and his/her representative.

4 Special interest groups which are bound together by their concerns, professional groups with lay representatives for questions of medical ethics, etc. pressure groups with commercial interests, and environmental groups with and without a political agenda.

5 How individuals can make up their minds and influence the state apparatus. This may include thinking about television and newspapers, the information they provide and the bias they may transmit, as well as the more general question of the public's 'right' to information, and rights in law.

That seems a formidable list for either teacher or student. Nevertheless it seems quite essential that those who are seriously teaching STS face up to these general dimensions of decision-making. They form neither just a context for science and technology, nor an add on ('and this is how a public inquiry operates') but are central to the whole concept of citizen science. Furthermore, they are every bit as controversial as is any other part. Social apparatus is no more static than is science and technology, so it follows that students may be invited to comment on the provision for public participation in decisions about new innovations, as well as to play their part in them.

Like other parts of STS which have been discussed in previous chapters, the nature of social decision-making is not best tackled in the abstract. No teacher is invited to work their way painfully down the list given above, point by point. As in the case of the *nature of science* it is topics being taught which supply the context in which the role of government, citizen or pressure group comes up for comment. Out of a number of such cxamples, some local, some political, and some global, STS teachers can ensure that their students are learning about the place of science and technology within the constraints of a democratic society.

Simulation for understanding social decision-making

One arena for decision-making, that of public inquiries, has certainly not been ignored by the writers of simulation resources. Usually they dissect the problem, which might be the siting of some potentially hazardous plant, in terms of the arguments which might be presented by different interest groups. Thus it falls neatly into a kind of team game in the classroom. The great advantage of this is that several important points of view are presented to the students, and information for the various teams can be provided. Once again, however, the criticisms which have been levelled against other simulations cannot be avoided. The students are marshalled into predetermined camps, the agenda is set, and committed students may find no way to express their own beliefs.

It is difficult, although not impossible, to avoid these traps while keeping to the simulation format. Simulations about current concerns are particularly successful. They both catch the students' interests and ensure that the fictitious 'gaming' aspect is avoided. It is also very valuable to simulate a procedure – inquiry, court of law, or inquest – at the time when public attention is focused upon it. The organisational problems include providing information (not difficult if there is plenty of newspaper and television coverage) and allowing space for individual, and possibly emotive, points of view which may be less easy.

The following account is from a teacher who set up an impromptu simulation of a court case in the wake of the trial of a doctor who had, at the parents' urging, let a severely handicapped baby die without treatment. The emotional story had hit the headlines for two weeks and was a natural subject for argument in the sixth-form common-room. The simulation allowed the teacher to show a little of the operation of a real court of law.

> We set up a trial complete with judge and jury. I appointed lawyers for each side, but it was the students who decided, with a little prompting from me, what the other roles should be. I had to suggest character witnesses, and then they took off. They had thought that only specialist witnesses were called, just doctors and the guy that did the autopsy. The great thing about character witnesses was that they could use their own experiences, you know? – didn't have to mug up a lot of info. And it was these characters which brought out the students' own feelings and values.
> The one I remember best, and this took place some years ago, was done by a girl of less than average ability who volunteered, rather shyly, to talk about having a severely handicapped child in the context of the court case. She must have worked up the idea in the next couple of days, or perhaps she already held strong views on the subject; I really don't know. Anyhow when the simulation was on and the prosecuting lawyer asked if she loved her handicapped son, and if she would let him be killed, she did just wonderfully. I can't remember the exact words but something like – 'Of course I love him now, but all the pain he's had, perhaps it would have been better if the doctors had not struggled so hard to keep him alive. No one ever asked me what I wanted'. You could tell her heart was in it!

Summary

This chapter has produced an odd assortment of evidence about simulations. Some of the examples have been given in the words of those who carried them out because there is a strong personal element in the choice of strategy. Science teachers vary enormously in their attitudes towards any sort of 'acting' in the classroom; some love it, others are frankly scared. If the context is unfamiliar to the class, the teachers' misgivings about the activity may be justified.

The internal dynamics and decisions of the industrial boardroom are by no means central to

STS, although they can be instructive if well done. Civic contexts for decision-making, such as courts of law and public enquiries are closer to the citizen and may produce rewarding simulations. The best of these strategies which (a) use a valuable context for the simulation, (b) which is adequately familiar, and (c) allows for the expression of individual views, can be quite excellent.

CHAPTER 7

Group discussion of issues – the DISS Project

Introduction

This chapter is devoted to the subject of group discussion of science-based social issues in the classroom. For the most part it uses the findings of the DISS (Discussion of Issues in School Science) project which is a very rich source of data on student talk. With its help we can begin to see the variety of ways in which students regard STS issues, and how they talk with each other about them. This is research data which is not only of the greatest interest in its own right, it also supports a way of teaching and learning which brings out students' value positions, relates these to our complex society, and considers possible citizen action.

There has been a fairly long tradition of classroom 'discussion' of controversial issues in the humanities and the social studies. At first it was used as an occasion for teaching the skills of debate. This had two dimensions. First it was considered to be important for students to be able to express themselves orally, and secondly to be able to marshal evidence in an even-handed way. Making personal value positions explicit was not part of this programme.

It was the Humanities Curriculum Project (HCP) of the 1960s which first took seriously the promotion of more genuine discussion. This was to go beyond the closed agenda of weighing up the evidence from both sides of the case and coming to the supposedly 'logical' conclusion. For the first time the HCP accepted that value judgements

were not only acceptable but might even transcend the paper data. It followed from this that the resulting decision-making could not but be a fallible process: there would be no 'right answer'. That immediately posed another problem – what was the teacher's role during the discussion? She or he could no longer direct the discussion towards the uniquely correct dénouement but it did seem important, in the mêlée of different value-positions, that the teachers did not try to intrude their own values. In such personal matters there should be absolutely no indoctrination. Out of this debate was born the notion of the 'neutral chairman' (Stenhouse, 1969). The teachers could certainly run these new types of discussion, but they should take care to give no inkling of where their own opinions lay.

Teachers tried to follow the recipe of the neutral chair, but it was difficult. In STS, or any other course devoted to the notion of encouraging concerned citizenship, it was particularly difficult and, in some sense, quite self-defeating. If the implicit message was that adults should become involved in public concerns then it was surely odd to find that the very teacher who had transmitted the message was pretending not to abide by it. It was inconsistent and embarrassing.

The next fashion in the conduct of discussion was the 'balanced chairperson', closely followed by the 'Devil's advocate'. It is all too easy to poke fun at the sober educational material which proposed one strategem after the other with so little evidence of having tried out any of them. There are certainly some occasions when either one or the other has its merits. Teachers who tried to introduce appropriate strategies for discussion learned the hard way that large class groups were not able to 'discuss' in any meaningful way, and that small friendship groups of students felt more free to talk about social issues without the direct participation of the teacher. This meant that the students had to arrange, lead and report on their discussions for themselves. Each group thus became a symbol of free speech within a society, listening to each other and reporting any agreements reached. It was an important landmark for the classroom, and for STS.

Public understanding of science

STS is basic to the public understanding of science both because it is an educational movement related to the connections between science, technology and society, but also because it emphasises those aspects of science which are most relevant to lay people, and pays attention to their values, their rights to information, and their capacity for action. The 1980s have, in some respects, been a decade devoted to exploring and expanding scientific literacy which is a much wider issue than some of the more simplistic reports of what questions of fact can be readily answered by the person in the street. In 1985 the British Royal Society brought out a report on *The Public Understanding of Science* (Bodmer) which made a number of specific recommendations for improving the nation's knowledge of science. Amongst its comments were some on the importance of school education, and others which deplored the sensationalist aspects of much media communication of science.

Following upon this report, the Science Policy Support Group set up a series of linked research projects (funded by the ESRC), designed to explore different facets of the public's understanding. Two of these were large-scale surveys (Durant *et al.*, 1989; Breakwell, 1990), several others were case studies which looked at small groups who had special need for particular scientific information (e.g. Wynne 1988, which recorded how Cumbrian farmers reconstructed the information that they

received from scientists in Whitehall concerning the radiation count of their lambs after the Chernobyl fallout). A third group was concerned with the fashioning of scientific communication to the public.

DISS was the only project to have a school setting, but it also shared some features with the others – it included a questionnaire about attitudes and interests, used media communication, and focused on small groups of individuals. The project took advantage of the STS courses already set up in schools which had small group discussion as a part of their internal assessment. From here it set out to explore how students used scientific knowledge when they discussed controversial issues which they had been watching on television.

The research reflected, in microcosm, what is probably the commonest way of getting to understand science-based issues for adults and students alike. What was studied was *public understanding in the making*, not only because of the young age of the participants, but also because, during the train of talk, it was possible to see opinions being clarified, exchanged and even formed from scratch. Studies of environmental issues among American high-school students had already indicated that television is the most frequently quoted source of information and there was no reason to suppose that the situation was different in Britain. Other data (McQuail, 1984; Hodge and Trip, 1986) reported that both children and adults seem to feel the need to discuss television programmes *in order to work out their understanding of them*.

Some students seem to acquire a large part of their information on most subjects through discussions with friends. Oral contributions are rarely recognised in the school assessment system but they were pioneered in STS examinations during the early 1980s. Now, in the context of research, it became possible to use recorded group discussion in the classroom, after watching television, as data for exploring how information is received, exchanged and reconstructed. In this way our research was to have a substantially different complexion from those which had been carried out before on students' attitudes and values; it would rely more upon data from *social interaction* than on *individual probes* like interviews and question-naires.

Group discussion in action

One of the aims of STS education was stated, in Chapter 1, as the discussion of personal opinion and values, as well as the participation in democratic action. In Chapter 6 some large-scale activities were mentioned which involved the expression of opinions and values, but these are not perfectly suited to all kinds of issue. In particular for shy students, for deeply felt issues, and for those students who have not yet worked out where they stand, supportive discussion in small groups may be just the right environment for this important facet of STS.

The philosopher David Bridges, in his book on *Education, Democracy and Discussion* argues that there are four functions for discussion when it is applied to a controversial issue. All useful talk begins with a working out and sharing of personal perspectives on the topic. This may sound rather trivial, like the endless superficial discussions of characters in last night's soap opera, which can be heard all round the school. Nevertheless it can be the outward sign of beginning an inner deliberation. How and where these discussions finish indicate different kinds and levels of achievement (Bridges, 1979, p. 44):

(a) For some the sharing of perspectives is a sufficient goal in itself.
(b) For others, who listen carefully, reaching an understanding of the variety of available responses is important.
(c) Those who are more reactive and responsive will go further into a stage where some kind of subjective choice between different values is made.
(d) If the participants can find a rational resolution to the controversy they will have reached a further stage of planning citizen action.

In stage (a) the expression of personal values requires time and a supportive environment, but should lead on naturally to stage (b). That is why a friendship grouping of students was recommended in the project so that the students could listen carefully and responsively to each other. At stage (c) there must be an element of 'weighing up alternatives' (Kitwood, 1984) which is the basic meaning for that over-used term 'evaluation'. This may be partly private but also social and sociable through the mechanism of trying out one's opinions on friends. Students who do not yet know where they stand are likely to go through a process of what sounds like hesitant deliberation. Achievement of the final stage (d) does not indicate that all the world's problems have been solved, only that there is a consensus on what has gone wrong and what 'right action' might look like. That implies that this stage is also a values achievement.

How the research took place

Fourteen secondary comprehensive schools in different locations helped with the work. Two were sixth-form colleges (age 16–19), and the others catered for the complete 11–19 age range. The students were not, for the most part, taking an 'A'-Level course. There was no way in which our sample could have been tailored to suit a research criteria, and the very fact that they had elected to take an STS unit gave the groups a special composition.

Research began by selecting a series of six video extracts to be shown in class. Each excerpt had its own special features:

1 The first was donation of kidneys for transplantation was taken as a *personal* issue by the students. It raised ethical and cultural points as it visited the USA and Japan, and discussed buying kidneys from live but needy donors.
2 The second video was on the more political issue of nuclear power – its risks and costs. Different government attitudes and public opinion were shown, in Britain, France, Sweden and the USA. The presentation was itself contentious and the video finished in the midst of a heated debate between the (then) Chairman of the

Central Electricity Generating Board and the organiser of Friends of the Earth.

3 A third video excerpt was about genetic counselling.

4 Compensation for veterans of the Atomic Tests which took place on Christmas Island in the early 1950s was another.

5 The fifth was access to information about industrial effluent.

6 The final video excerpt concerned Third World medicine showing efforts to combat blindness due to vitamin A deficiency in parts of Africa and India, and finishing in a mobile eye clinic for cataract operations.

Group discussion is usually managed by the teacher to a set agenda of questions. However in this work we wanted to explore the student's own agenda – what they found to be important. So we urged teachers to ask no general questions but simply to encourage their students to form groups, talk together, and record their own discussions with the tape recorders which we supplied.

The six television excerpts were from general, not educational, programmes. They had all the usual characteristics of media communication, emphasising the worrying features, and assuming little previous knowledge. Several of them were also, in terms of the Bodmer Report, somewhat sensationalised. Each video extract lasted no more than 20 minutes. They were shown in the same order during the year, but at different times to suit the individual teachers. The students' discussions which followed were usually about 10 minutes long – although there were cases where they went on for half an hour, and a few (about 8 per cent) where the students seemed unable to begin at all. By the end of a busy two years, however, we were in possession of about 200 tape recordings, all of which we had carefully transcribed, of young people talking together with all the frankness and seriousness that we could have desired.

Three functions of talk

One common mode of discussing simply made reference to incidents from the video. At first sight this seemed totally redundant in view of the fact that all the students had been similarly placed to receive the excerpt. Nevertheless it was clearly important for them to be sure that others had noted what they had noted, and that what they found impressive, surprising, or distressing, was found to be the same by others. Sometimes items of science knowledge were included in this phase of talk for confirmation about what had been said and the way it should be understood. 'There was radioactive stuff in the smoke, wasn't there?', or 'The chance was one in 25, yeh?' In this way the reception process began to set an agenda for subsequent discussion by identifying important issues. We called this kind of talk *framing*.

The second type of talk was less tentative and more deliberative. Personal opinions and ideas were tried out and exchanged so that the speaker could begin to clarify his/her own views, and also to see if the others might agree and reinforce them.

'. . . and life's not going to be that good for the child so maybe it's the best decision, don't you think?' or 'I couldn't cope with a handicapped kid, could you?'

This was the stage of discussion at which *collaborative speech*, where one student completed a sentence begun by another, was most common. Frequently questions were posed in a ruminative way, to see how some possibility might work out. 'What would you do if it was your mother?' was a favourite gambit of this kind. Items of scientific or medical knowledge were treated in the same questioning and rhetorical way: 'Why do we have two kidneys then if we only need one?'

The third mode of discussion was more argumentative and committed. Sometimes the students even counted up who was 'for' and 'against' in the manner of an election. As they spoke they openly persuaded, rather than merely deliberated, and they did this in one of several rhetorical ways. A favourite method was simply to exemplify the issue – 'My father has had three unsuccessful operations like that.' Alternatively they posed a personal or intimate situation followed by a question:

'Suppose they built a nuclear power station in *your* garden?'

'How would *you* like *your* kidney to be in a murderer?'

The third rhetorical device involved putting speech into the mouth of another.

'They would say "Don't worry, just give us your kidney, you'll be all right",' or

'So you would say to your wife "Shame about our child's leukaemia, I want employment!" '

Three points must be borne in mind about this classification. In the first place, while it is true that the forms of logic, 'if . . . then . . .' were vanishingly rare, it does not follow that the talk was irrational. Secondly, the three types of talk are by no means always easy to distinguish. Indeed our early attempts to identify and quantify the three different types of talk on every transcript met with such poor success in terms of inter-researcher reliability that it had to be abandoned. A proportion of the discussion passages were very clearly of one kind or another, but there were too many cases where it is difficult to distinguish between deliberation and commitment, or between personal clarification of values and deliberation.

The third point concerns the order of these types of talk. It was not the case that one followed on the other, as logic might seem to dictate. Discussion could begin with a committed outburst against some aspect of the issue, only to be followed by a quieter passage of social deliberation. Other groups began with deliberation and then went back to framing talk to gather more evidence from their memory of the video.

In every case, however, any ideas for strategies to deal with the problems under consideration were discussed during the committed type of talk. For this purpose more knowledge was sometimes required and students asked about or volunteered more information. Sometimes this was of a scientific kind, like the effect on effluent distribution of raising the height of a factory chimney, or legal and social such as possible legislation or the operation of trust funds.

Scientific knowledge?

Reading through the discussion transcripts gave the first impression that little or no school science knowledge was being used, just as most other researchers on STS have reported. A closer examination suggested, however, that familiarity with simple science terms and concepts was underpinning every bit of the discussion work. It was only when this knowledge failed, when the easy understanding broke down and students used the old jibes 'Why don't they speak English!' or 'They try to blind you with science', that we realised how much they had been relying upon *familiarity* with science terminology. This may not be equivalent to orthodox definition but it served to allow meaningful talk to continue. As Wynne (1988) has remarked in another context, public understanding makes knowledge about science 'invisible' in normal discourse. Only when this familiarity failed did terms like 'catalytic converter' or 'radioactive half-life' become a matter for bewildered comment.

Finding out how much scientific knowledge was being used was harder than we expected. In the first place explicit tutoring and mini-lectures on scientific concepts were rare events for reasons which, we suppose, are more related to adolescent norms of behaviour than ignorance. This made identifying the depth of understanding, or its source, quite difficult and uncertain. Secondly, personal prejudice or history brought about a kind of mental filtering which could distort the information given on the video. It was common, for example, for Japanese to be reported as Chinese, or for Third World doctors to be thought of as no more than health volunteers. Thirdly, experiential information about ethnic groups, individual action, or civic strategies were such essential facets of knowledge that it was difficult to know where to divide scientific from social information. This interweaving of different kinds of knowledge has also been a feature of some other studies of public understanding. Nevertheless, we did score the students each time they added a piece of scientific knowledge to the discussion, just as we scored references to television programmes on science, to reading about science, and claims that more information about science was needed.

Discussion behaviour and questionnaire claims

In an effort to wring some more generalisable conclusions from the data, we constructed a profile of all the items of discussion behaviour which we were able to record with an inter-researcher reliability of better than 90 per cent. (These were only compiled for the 112 students who had taken part in at least four discussions.)

Pre- and post-course questionnaires were administered with five main sections exploring the following general areas:

1 School experience and aspirations in science.
2 The origin of their knowledge about 'science issues'.
3 Attitudes towards particular issues.
4 Understanding of the nature of scientific knowledge.
5 Views on civic responsibility for science issues.

The purpose of this data collection was not simply to record the percentages in each area, but to make a comparison between how the students spoke during the discussions, and what each student had claimed in section II of the questionnaire.

The majority of the students who had taken part in the discussions agreed that: 'My science teachers have often mentioned science issues during lessons.' This was hardly surprising in view of the STS course they were following, but it effectively eliminated this from our investigations. The three other categories of knowledge sources claimed in the questionnaire responses remained:

(a) knowledge from TV
(b) knowledge from talking with friends
(c) knowledge from reading

The slightness of the associations between what the students wrote in their questionnaires and what they said in their discussions was striking. The associations between claiming that most of their knowledge came from television

and mentioning television programmes during discussion, between claiming that they got quite a lot of information from reading
and mentioning reading during discussion, and between claiming that most of their knowledge came from talking with friends
and being persistent during discussion

were either non-existent or not significant.

The lack of expected associations between these two types of score was repeated in the section of the questionnaire which concerned civic responsibility (section V). Thus the associations between agreeing with the statement about taking individual action or joining a pressure group had only weak associations with either mentioning individual action or civic strategies during discussions.

This is far from being the only set of research data where expressions of attitude in a questionnaire have failed to show any correlation with actual behaviour. It does not necessarily mean that the questions were at fault. It could simply mean that the students 'presented themselves' through the questionnaire in a way they wanted to be recognised. A factor analysis of the questionnaire data showed clusters of responses describing student 'types' (including those related to gender) which are very familiar. The major factors arising from preferred questionnaire responses for boys and girls are presented below in a condensed form.

Even the data about preferred knowledge sources showed some features which have been reported elsewhere. There has been research within schools from the USA (Iozzi, 1984) and

Girls	Boys
When certain I try to convince others	Those who worry about issues are probably ignorant
Most of my knowledge about issues comes from TV	My science teachers do not mention issues
My views on issues are probably like my friends'	Issues are best left to the scientists
More likely to watch TV than go out with friends	My views on issues are probably like my friends'
My views on issues are probably like my parents'	Issues mentioned are all good for society

from national educational initiatives in South America and Africa (reviewed in Greenfield, 1984) suggesting that there may be very little overlap between groups who favour information reception via watching and via reading, although there is much more in common between those who favour watching and those who discuss with friends. This matches nicely with our finding that those who claimed to get most of their knowledge from television, and those who claimed to get most from talking with friends, shared other features like talking in a collaborative fashion and not mentioning any intention to take individual action. 'Readers' on the other hand, had a negative association with mentioning TV, just as 'talkers with friends' had with reading.

On the basis of these data we might conclude that the three groups (designated by claimed knowledge sources) were showing how they thought of themselves in relation to knowledge, rather than expressing any *information-processing preference*.

Do boys and girls talk differently about moral issues?

There has been work which suggested that there is a profound difference in the ways in which the two sexes address moral questions (Kohlberg, 1984). The discussion profiles of boys and girls showed no significant difference on practical suggestions for social action, either at an individual or a civic basis. The transcripts showed that different topics produced very different types of response, e.g. individualistic on kidney donation, political on nuclear power, gender-related on genetic counselling. This meant that we had to select just one topic for closer investigation.

We examined the 14 transcribed discussions which had been recorded during the first year of the project, about the veterans of the Atomic Bomb tests. From these three different categories of moral response were scored which might resemble those identified by Kohlberg.

'Broad statements'
These comment upon the problem or issue from an over-arching perspective. Kohlberg defined his sixth, and highest stage of moral development as 'universal principles of justice' (Kohlberg, 1984, pp. 174–77). Occasional statements by the pupils were decontextualised in this way: 'It's the principle not the money.' However the majority of the students' judgemental statements, in which they stood back from the details of the situation, focused more specifically on what the government should do, on the servicemen's rights, or on regulations which should be enforced:

> 'They should be allowed compensation because it was not voluntary.'
> 'Yeah, I mean that isn't right. I don't think it's right. You shouldn't have to pay with your lives.'

'Personal statements'
At the opposite extreme is a type of comment which shows the speakers so strongly empathising with the people in the video that they try to include themselves in the situation. They attempt to understand the issues by imagining how *they* would feel, and what *they* would do if directly involved. Kohlberg placed such statements – 'putting yourself into other people's shoes' – down at stage 3 in moral development. We included in this category statements where the speakers encouraged others to imagine themselves in the situation.

> 'I'd rather die first if they dropped the bomb anyway – I'd go and sit myself underneath it.'
> 'Would *you* worry?'

'Contextual statements'
This third category is characterised by comments which are neither abstract, nor yet so completely personalised as those in the second category. Here the speakers take account of the situations of the people and events in the video, and add comments based on their own experience or understanding of similar situations. Some showed strong emotive reactions, others were cooler and even cynical, and many simply recalled factual details from the video which seemed relevant to the present stage of the discussion.

> 'The children are the ones that are going to live with it.'

'I think it could be because hc [the official] is probably in favour of the government, and he could be a bit biased.'

Results of the gender analysis

1 *By far the largest number of contributions from both girls and boys were contextual.* The students seemed to be using their talk to get a grasp of the situation and needed to mull over the details of the story in order to underline those issues which seemed salient (similar to 'framing' talk in the earlier categorisation).

2 The proportions of broad to other comments made by each boy and girl in the Atomic Test discussions was calculated for the mixed gender group discussions. *Both boys and girls seemed to make broad statements during discussion more in accordance with individual or group styles of argument than with gender.*

3 *The opening statements of the discussions seemed to serve special managerial purposes.* In one way or another the students – girls or boys – seemed to be attempting to get the discussion going. There were several tapes showing a hesitant start with words like 'Right . . .' or 'Well . . .'. Others, in the manner of a chairman, began almost formally with 'We're talking about test veterans . . .' (see also Barnes and Todd, 1977).

4 In every discussion which opened with a broad statement, this was made by a male student. This finding was corroborated by a search through the 93 transcriptions of the different discussions held the first year. The results of this exploration confirmed that *where the opening statement was in the broad category it was significantly more often made by a boy than by a girl* (Figure 7.1).

5 Using the same method as for (2), we found that *personal statements were* not *made more often by girls than by boys in the main sequences of discussion.*

6 Those statements which included reflections on civic action for dealing with the issue, e.g. about legal constraints or procedures which were thought to guard against future risk, or to compensate for an injustice appeared as a result of a thought out position.

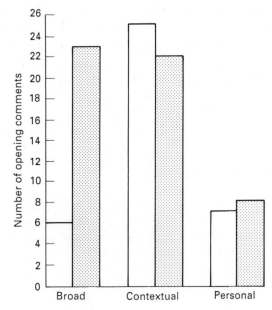

Fig. 7.1 Opening comments (*n*=91): □ female; ⊠ male

e.g. '. . . there should be some laws . . .'
'You can't put a price on that . . . they should put [the money] into a big fund which you take out as you need it.'

In no case did these strategic statements follow an opening broad comment.

So it seemed that boys and girls did *not* differ much in the kinds of moral statements that they made. Only when they opened a discussion was there a clear difference of style. Boys tended to begin with a moral judgement which may have been somewhat 'off-the-cuff'; girls were more hesitant and deliberating.

Our results may have been dissimilar to others which had reported large differences between the sexes, because the DISS project gave data drawn from behaviour in a social situation. The basic purpose of social talk is to try out opinions, receive feedback, and respond to others. This necessitates taking on board others' perspectives and even, perhaps, their style of discourse. The questions and comments of friends in this setting may well override personal and gender-related styles and orientations. That in itself implies a recommendation for teachers to run their classroom discussions in mixed-gender groups.

STS, as we have argued, denies that science is value-free, and maintains that technology affects different sections of society in different ways. For both these reasons we are bound to listen to, and consider, the views of all involved in the issues of our times. This is an important point in our STS education. All matters of social justice need to be considered from points of view, different from our own. The philosopher John Rawls (1971) has written that we should consider what is best for our society as though we were hidden 'behind a veil of ignorance' about our own situation. However difficult that may be to achieve in practice, the moral analysis of the discussions given above suggests that some progress towards this goal was being made.

Change of attitudes on issues?

One section on the pre- and post-test questionnaire asked about the students' attitudes towards a range of issues, including those which figured in the discussions, on a five-point attitude scale. It seemed important to find out whether these responses showed a change in attitude over the year. What made this difficult was the undeniable fact that during the course of the year other factors, quite unrelated to the set-piece research discussions, must also have been responsible for substantial changes in the students' knowledge.

The recommended way to control for such factors is to collect data from another similar group which had not gone through the discussions. We had asked each school to find a control group to compare with those who had taken the STS course, but events showed that they did not, and indeed could not, be effectively matched. Even at the beginning of the year this was so. People are not tailor's dummies in respect to attitudes on issues. It was hardly surprising to find gross differences between the views of the two groups on social issues in view of the courses which they had chosen to take.

On the question of stopping all experiments on animals, which was *not* included in the discussion topics, both groups changed their attitudes, and in the same direction. On those issues which were discussed in the project, attitudes sometimes moved in different directions for the two groups (Figure 7.2). However it is still necessary to counsel caution in attributing changes of attitude unequivocally to the single discussion during the year which was held on the particular issue. The charts of Figure 7.2 also hide a considerable change of individual attitude. Such variations might be taken to show that no indoctrination was taking place during the STS lessons and that the students' opinions on some matters were quite fluid and changeable. But it also urges caution in claiming that any overall change in attitude on an issue was the direct result of class discussions alone. (It seems distinctly reductive to expect a free and serious exchange of views within a group of four friends, to produce a result which could be recorded unequivocally by a tick in one of five boxes.)

Any re-appraisal of personal values, or a new understanding about some important issue, might be expected to stimulate student thinking even after the talk has stopped. It could be argued that the 'speaker–hearers', as Barnes and Todd (1977) have called them, will have the argument continuing in their heads, and possibly in their talk, for some time after the discussion is over, if it has sufficiently engaged their attention. Conversation with the teachers showed us, during the course of the first year, that the topics discussed earlier were quite often referred to a later date when they were triggered by some item in the coursework or on the news. We asked a few of the teachers to collect short pieces of writing from the individual students about a week after the video discussion had been held. These were little more than half a page in length and some of them were strikingly similar, in choice of words as well as general argument, to the contributions made to the actual discussions. Such exact recall of what had been argued more than a week before provided unexpected evidence for continued reflection by the students on the class discussion and its arguments.

Personal values and civic responsibility

The students' attitudes towards civic responsibility

Fig. 7.2

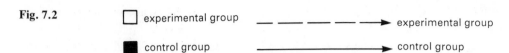

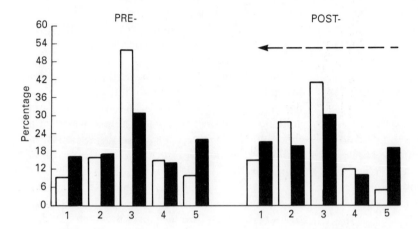

(a) Genetic counselling and abortion
(Discussed by experimental group)

(b) Building more nuclear power stations
(Discussed by experimental group)

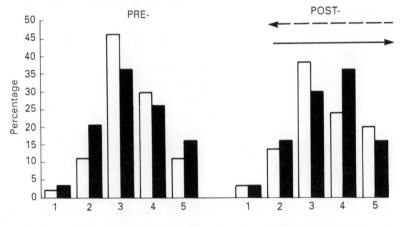

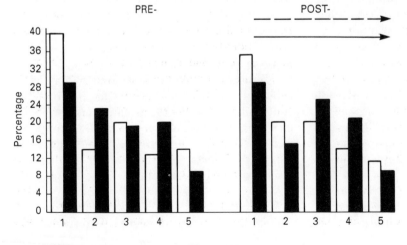

(c) Stopping all animal experiments
(Not discussed)

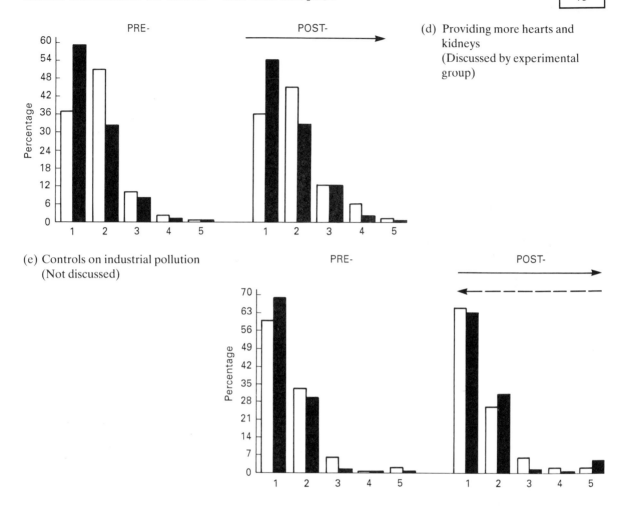

(d) Providing more hearts and kidneys
(Discussed by experimental group)

(e) Controls on industrial pollution
(Not discussed)

was yet another matter which was mixed up with the reception of knowledge and the construction of understanding. It has already been reported that the students' reception of information was coloured by their preconceptions: it was also affected by their trust in the figures of authority – the scientist, doctor, industrialist or politician – presented on the programmes. Often these were the very figures who might be thought to bear all the civic responsibility for decision-making on these issues. How much did the students trust them, or seem content to leave decisions to them? The discussion transcripts showed the students using social caricatures – 'All politicians would say that!' or 'They always lie' in their comments on the programmes. Caricatures, however, form poor guides to confidence in others' judgement, or to any wish for personal involvement and action.

It was surprising, perhaps, to find that the questionnaire responses did show a few significant correlations between academic achievement in the sciences (e.g. the number of sciences taken at GCSE, or the number of C grades or above which had been gained) and beliefs about civic responsibility for science issues; 86 per cent of the discussion group were taking a GCSE course with no A level subjects, while all the control group were taking an A level course in the sciences. Thus a comparison between these groups in the pre-course test is effectively one between groups at the same age with an interest in science (scored in the questionnaire at over 87 per cent) but markedly

Table 7.1

Statement	More agreement with statements in pre-test
1 'Deciding on science issues is the responsibility of government'	Discussion Group*
2 'Deciding on science issues is the responsibility of scientists'	Control Group**
3 'The politicians' main responsibility is to protect and encourage industry'	Discussion Group
4 'Science-based issues are too complicated for ordinary citizens to understand properly'	Discussion Group
5 'Once I am over 18 I shall vote in every election'	Control Group
6 'Really caring about issues means joining a group and doing something about it yourself'	Discussion Group*
7 'Scientific knowledge should be kept secret if it is industrially valuable'	Discussion Group

* Move of discussion group significant at $p < 0.005$).   ** Move of control group significant at $p < 0.05$).

different academic achievement in science which had led them into different sixth-form courses, and given them different career expectations.

Just after the GCSE examination results had come out and the students started their new courses, there were more differences between the groups than at the end of the year in the sixth form. This was an unexpected result (Table 7.1).

By the end of the year all except two of these differences had disappeared. The control group, with its better background of scientific knowledge, still believed more strongly that deciding was a responsibility of scientists, and also that the issues were too complicated for ordinary citizens.

All the very significant changes during the year had been made by the group which had been following the STS course and taking part in the DISS project discussions.

Evaluating STS

Of course it is not possible to claim that these changes of attitude towards civic responsibility were due to the discussions in particular, or to the STS course in general. They could just as well be a result of a year's maturation in the more adult atmosphere of the sixth form. At all events it was good to find that the STS group were no longer so keen to leave decisions to the government or to scientists. It was somewhat less satisfactory to find

that the group was now less inclined to believe that they would vote in every election than they were before the course started. This result is quite general, however, for the control group in this study, and in other research data. As the date of enfranchise approaches the appeal of voting seems to decrease.

Perhaps the greatest success recorded in this part of the attitude test, in terms of STS objectives, was that the group who had taken the course were now confident enough to disagree with the statement that these issues were too complicated for ordinary citizens.

On this note we may end this exploration of STS education with the words of a student of average ability who had just completed the year of study.

I learnt more about things I had just briefly heard of such as the creation of the Atomic bomb and how it affected the lives of millions. I studied and discussed the topics I enjoyed – the Third World, its health food and population. I have become more aware of how I could help in areas of interest.

By doing my project I was able to find out more about the charities which are helping with Third World health problems, such as Oxfam.

Overall I gained a lot of knowledge on science and technology which I hadn't heard of before.

It has made me more aware of how society differs on topics like nuclear power. It has made me respect other people's opinions.

References

Adey, P. and Shayer, M. (1990) Accelerating the development of formal thinking in middle and high school pupils. *Journal of Research in Science Teaching*, 27(3), 267–85.

Aikenhead, G. (1975) *Science: A Way of Knowing*. University of Saskatchewan, Saskatoon.

Aikenhead, G. (1989) High school graduates' beliefs about science technology and society. III. Characteristics and limitations of scientific knowledge. *Science Education*, 71(4), 459–87.

Aikenhead, G., Fleming, R. and Ryan, A. (1987) High school graduates' beliefs about science technology and society. Methods and issues in monitoring student views. *Science Education*, 71(4), 459–87.

Association for Science Education (1988) *Minerals in Buenafortuna*. ASE, Hatfield.

Bacon, F. (1605) Kitchin, G. W. (ed.) (1973) *The Advancement of Learning*. Dent & Sons, London.

Barnes, D. and Todd, F. (1977) *Communication and Learning in Small Groups*. Routledge, London.

Bodmer, W. (1985) *The Public Understanding of Science*. Report of the Royal Society, London.

Boeker, E. (1979) *Science, Society and Education*. Vrije University, Amsterdam.

Bonati, G. and Hawes, H. (eds) (1992) *Child to Child*. Child to Child Trust, London.

Breakwell, G. (1990) Young people's attitudes to scientific change. Paper given at the Public Understanding of Science Conference. Science Museum, London.

Breakwell, G., Fife-Schaw, C., Lee, T. and Spencer, J. (1987) Occupational aspirations and attitudes to new technology. *Journal of Occupational Psychology*, 60, 169–72.

Bridges, D. (1979) *Education, Democracy and Discussion*, NFER, Windsor.

Browne, N. and Ross, C. (1990) Girls' stuff. Boys' stuff: Young children talking and playing. In N. Browne (ed.) *Science and Technology in the Early Years*. Open University Press, Milton Keynes.

Collings, J. and Smithers, A. (1984) Personal orientation and science choice. *European Journal of Science Education*, (6), 295–300.

Davies, F. and Greene, T. (1984) *Reading for Learning in the Sciences*. Oliver & Boyd, Edinburgh.

Driver, R. and Oldham, V. (1986) A constructivist approach to curriculum development in science. *Studies in Science Education*, 13, 105–22.

Durant, J., Evans, G. and Thomas, G. (1989) The public understanding of science. *Nature*, (340), 11–14.

Ellington, H., Addinall, E. and Percival, F. (1981) *Games and Simulation in Science Education*. Kogan Page, London.

Furth, H. (1980) *The World of Grown-ups. Children's Conceptions of Society*. Elsevier, New York.

Gould, S. J. (1990) *Hen's Teeth and Horse's Toes*. Norton, New York.

Greenfield, P. (1984) *Mind and Media*. Fontana, London.

Hartup, W. (1978) Children and their friends. In H. McGurk (ed.) *Issues in Childhood Social Development*. Methuen, London.

Head, J. (1983) *The Personal Response to Science*. Cambridge University Press.

Hodge, B. and Tripp, D. (1986) *Children and Television*. Polity Press, Cambridge.

Hogben, L. (1938) *Science for the Citizen*. Unwin, Woking.

Iozzi, L. (1984) *Summary of Research in Environmental Education*. Monographs in Environmental Education and Environmental Studies, ERIC Clearinghouse for Science, Mathematics and Environmental Education. Ohio State University, Columbus.

Jegede, O. (1991) The relationship between African traditional cosmology and students' acquisition of a science process skill. *International Journal of Science Education*, 13(1), 37–47.

Kitwood, T. (1984) Cognition and emotion in the psychology of human values. *Oxford Review of Education*, 10(3), 293–302.

Kohlberg, L. (1984) *Child Psychology and Childhood Education*. Longman, New York.

Layton, D. (1988) Revaluing the T in STS. *International Journal of Science Education*, 10(4), 367–79.

McConnell, M. (1982) Teaching about Science Technology and Society at the secondary school level in the United States. An educational dilemma for the 1980s. *Studies in Science Education*, (9), 1–32.

McQuail, D. (1984) *Communication*. Longman Group, Harlow.

Mead, M. and Wolfenstein, M. (1955) *Childhood in Contemporary Culture*. University of Chicago Press.

Mead, M. and Metrau, R. (1957) The image of the scientist amongst high-school students. In B. Barber and W. Hirsch (eds) *The Sociology of Science*. Macmillan, New York.

Meadows, D. *et al.* (1972) *The Limits to Growth*. Potomac Associates, Washington.

Medvedev, Z. (1969) *The Rise and Fall of T. D. Lysenko*. Columbia University Press, New York.

Meyrowitz, J. (1984) The adultlike child and the childlike adult: socialisation in an electronic age. *Daedalus*, (Summer), 19–48.

Musgrove, F. and Taylor, P. (1969) *Society and the Teacher's Role*. Routledge & Kegan Paul, London.

Peterson, A. D. C. (1973) *The Future of the Sixth Form*. Routledge & Kegan Paul, London.

Rawls, J. (1971) *A Theory of Justice*. Harvard University Press, Cambridge, Mass.

Reid, W. and Filby, J. (1982) *The Sixth: An Essay in Education and Democracy*. Falmer Press, Lewes.

SATIS 16–19 (1991) *The Framework Units* (Hunt, A. and Solomon, J.). Association for Science Education, Hatfield.

Shayer, M. and Adey, P. (1981) *Towards a Science of Science Teaching*. Heinemann, London.

Skilbeck, M. (1984) *School-Based Curriculum Development*. Harper, London.

Solomon, J. (1983) *SISCON-in-Schools*. Basil Blackwell and ASE, Hatfield.

Solomon, J. (1988) Science technology and society courses: tools for thinking about social issues. *International Journal of Science Education*, 10(4), 379–87.

Solomon, J. (1991a) *Exploring the Nature of Science*. Blackie & Sons, Glasgow.

Solomon, J. (1991b) Science investigations as a school-home link. *Links*, 17(3), 5–8.

Solomon, J., Duveen, J., Scott, L. and McCarthy, S. (1992) Teaching about the nature of science through history: action research in the classroom. *Journal of Research in Science Teaching*, 29(4), 409–21.

Stenhouse, L. (1969) The nature and interpretation of evidence. In *Authority, Education and Emancipation*, pp. 115–19. Heinemann, London.

Tizard, B. (1986) The impact of the nuclear threat on children's development. In M. Richards and P. Light (eds) *Children of Social Worlds*. Polity Press, Cambridge.

Werskey, G. (1978) *The Visible College*. Penguin Books, London.

Williams, A. (1990) Children's pictures of scientists. Paper presented at a conference on Publics and Policies for Science. Science Museum, London.

Wynne, B. (1988) Sheep farmers and scientists. Paper given at the Public Understanding of Science Symposium. Lancaster University.

Yager, R. (1980) *Analysis of Current Accomplishments and Needs in Science Education*. ERIC/SMEAC Clearinghouse for Science, Mathematics and Environmental Education. Ohio State University, Columbus.

Young, M. (1971) An approach to the study of curricula as socially organised knowledge. In M. Young (ed.) *Knowledge and Control*. Collier-Macmillan, London.

Ziman, J. (1980) *Teaching and Learning about Science and Society*. Cambridge University Press.

Index